Productivity and its determinants are at the heart of economic debate. Output per person or per capita is still the most influential measure of the prosperity of nations. Productivity depends on the quantity and quality of the factors of production available to a country and the social framework within which they operate. Education and the research base affect both the quality of factors and the ability of a nation to produce. This volume brings together papers from a number of authors from a variety of traditions. The importance of the growth and measurement of service productivity are addressed. The role of human capital in adapting to new technologies is discussed. The creation of knowledge through research and development and its diffusion through trade, investment and the interaction of firms are fully investigated. The volume starts with a discussion of differences in productivity between nations, and provides a comprehensive set of discussions as to why they exist.

RAY BARRELL is a Senior Research Fellow at the National Institute of Economic and Social Research, a visiting professor at Imperial College London and a part-time professor at the European University Institute in Florence. He has written widely on European economic integration, labour markets, and on the processes affecting the pattern of trade and investment. Previous books include *The UK Labour Market: Comparative Aspects and Institutional Developments* (Cambridge, 1994) and (with Nigel Pain) *Innovation, Investment and the Diffusion of Technology in Europe* (Cambridge, 1999).

GEOFF MASON is a Research Fellow at the National Institute of Economic and Social Research. His main research interests are in the field of comparative productivity performance and human capital formation; in particular, the analysis of labour markets for high-level and intermediate skills. He has published widely on these areas in Britain and the United States.

MARY O'MAHONY is a Research Fellow at the National Institute of Economic and Social Research. She has published widely in the areas of international productivity and capital stock measurement and the importance of physical capital in explaining the postwar performance of British industry and services. Recently she has extended her research on productivity to include the effects of intangible investment and regulation. Her recent book on Britain's relative productivity performance, published by NIESR, has been widely cited in the media.

THE NATIONAL INSTITUTE OF ECONOMIC AND SOCIAL RESEARCH

OFFICERS OF THE INSTITUTE

PRESIDENT
SIR BRIAN CORBY

COUNCIL OF MANAGEMENT
JOHN FLEMMING CHAIRMAN

PROFESSOR C. BEAN	D. JULIUS
I. BYATT	SIR STANLEY KALMS
SIR DOMINIC CADBURY	R. KELLY
F. CAIRNCROSS	H. H. LIESNER
SIR BRIAN CORBY	SIR PETER MIDDLETON
SIR JOHN CRAVEN	J. MONKS
J. S. FLEMMING	PROFESSOR N. STERN
PROFESSOR C. A. E. GOODHART	PROFESSOR K. F. WALLIS
LORD HASKINS	M. WEALE

DIRECTOR
MARTIN WEALE

SECRETARY
FRANCIS TERRY

2 DEAN TRENCH STREET, SMITH SQUARE, LONDON, SW1P 3HE

The National Institute of Economic and Social Research is an independent, non-profit-making body, founded in 1938. It has as its aim the promotion of realistic research, particularly in the field of economics. It conducts research by its own research staff and in cooperation with universities and other academic bodies.

Productivity, innovation and economic performance

Edited by

RAY BARRELL, GEOFF MASON and MARY O'MAHONY

CAMBRIDGE
UNIVERSITY PRESS

PUBLISHED BY THE PRESS SYNDICATE OF THE UNIVERSITY OF CAMBRIDGE
The Pitt Building, Trumpington Street, Cambridge, United Kingdom

CAMBRIDGE UNIVERSITY PRESS
The Edinburgh Building, Cambridge CB2 2RU, UK   http://www.cup.cam.ac.uk
40 West 20th Street, New York, NY 10011–4211, USA http://www.cup.org
10 Stamford Road, Oakleigh, Melbourne 3166, Australia
Ruiz de Alarcón 13, 28014 Madrid, Spain

© The National Institute of Economic and Social Research 2000

This book is in copyright. Subject to statutory exception
and to the provisions of relevant collective licensing agreements,
no reproduction of any part may take place without
the written permission of Cambridge University Press.

First published 2000

A catalogue record for this book is available from the British Library

ISBN 0 521 78031 4 hardback

Transferred to digital printing 2003

Contents

List of contributors	*page*	vii
1	Introduction RAY BARRELL, GEOFF MASON and MARY O'MAHONY	1
2	Britain's productivity performance in international perspective, 1870–1990: a sectoral analysis STEPHEN BROADBERRY	38
3	Labour productivity and product quality: their growth and inter-industry transmission in the UK 1979–1990 CHRISTINE GREENHALGH and MARY GREGORY	58
4	Productivity, employment and the 'IT paradox': evidence from financial services MICHELLE HAYNES and STEVE THOMPSON	93
5	Productivity and service quality in banking: commercial lending in Britain, the United States and Germany GEOFF MASON, BRENT KELTNER and KARIN WAGNER	117
6	Productivity growth in an open economy: the experience of the UK GAVIN CAMERON, JAMES PROUDMAN and STEPHEN REDDING	149
7	Innovation and market value BRONWYN H. HALL	177

8 Gross investment and technological change:
   a diffusion based approach   199
   PAUL STONEMAN and MYUNG-JOONG KWON

9 National systems of innovation under strain:
   the internationalisation of corporate R&D   217
   PARI PATEL and KEITH PAVITT

10 Agglomeration economies, technology spillovers
   and company productivity growth   236
   PAUL GEROSKI, IAN SMALL and CHRISTOPHER
   WALTERS

11 Human capital, investment and innovation:
   what are the connections?   268
   STEPHEN NICKELL and DAPHNE NICOLITSAS

Contributors

Ray Barrell *National Institute of Economic and Social Research, London*
Stephen Broadberry *University of Warwick*
Gavin Cameron *Nuffield College, Oxford*
Paul Geroski *London Business School*
Christine Greenhalgh *St Peter's College, Oxford and Institute of Economics and Statistics, Oxford*
Mary Gregory *St Hilda's College, Oxford and Institute of Economics and Statistics, Oxford*
Bronwyn H. Hall *Nuffield College, Oxford, UC Berkeley, NBER and Institute for Fiscal Studies, London*
Michelle Haynes *University of Leicester*
Brent Keltner *RAND Corporation, Santa Monica, California*
Myung-Joong Kwon *Yonsei University, Korea*
Geoff Mason *National Institute of Economic and Social Research, London*
Stephen Nickell *Institute of Economics and Statistics, Oxford*
Daphne Nicolitsas *Institute of Economics and Statistics, Oxford*
Mary O'Mahony *National Institute of Economic and Social Research, London*
Pari Patel *Science Policy Research Unit, University of Sussex*
Keith Pavitt *Science Policy Research Unit, University of Sussex*
James Proudman *Bank of England*
Stephen Redding *Bank of England, New College, Oxford and CEPR*
Ian Small *London Business School*
Paul Stoneman *University of Warwick*
Steve Thompson *University of Leicester*
Karin Wagner *Fachhochschule für Technik und Wirtschaft, Berlin*
Christopher Walters *London Business School*

1 Introduction

RAY BARRELL, GEOFF MASON AND MARY O'MAHONY[1]

In describing the course of standards of living in the postwar world there are a number of undisputed findings. First, using GDP per capita as the measure of well-being, the United States has consistently appeared at the top end of the cross-country ranking. Secondly, growth rates of GDP per head in the US appear to be relatively low, reflecting postwar convergence of many countries to standards of living in the US. Thirdly, the process of convergence is far from being universal and a significant number of countries have failed to converge on the US.

These findings are illustrated in table 1.1, which shows per capita GDP for the US and selected countries in Europe, contrasting these with the position in some Asian countries. Thus in 1996 GDP per capita in the US was over five times that in China and more than ten times that in the Philippines. Growth rates of GDP per capita show the US experiencing low growth in the period 1950–73 relative to European countries and low growth rates since 1973 relative to many Asian countries. But the country with the lowest per capita income in this group, the Philippines, also experienced the lowest growth rates in the two periods.

Focusing on the huge disparities of experience between industrialised and developing countries, however, should not divert attention from the fact that even within the former group experience has also been varied and cannot be explained entirely by the convergence model. However, as table 1.1 shows, the US continues to lead the major European countries in terms of GDP per head by a wide margin. In 1996 GDP per head in the US was over 20 per cent greater than in West Germany or France and more than 35 per cent higher than in Britain or Italy. From 1950 to 1973 growth in GDP per head was in general

Table 1.1 *Recent levels and post-war growth of real GDP/person in Europe, USA and Asia*

	GDP/head, 1996 $1990 int[a]	Growth (% p.a.) 1950–73	Growth (% p.a.) 1973–96
United States	23,719	2.4	1.6
Norway	22,256	3.2	3.4
Denmark	19,803	3.1	1.7
West Germany	19,622	5.0	1.8
Netherlands	18,504	3.4	1.6
France	18,207	4.0	1.5
UK	17,326	2.5	1.6
Italy	16,814	5.0	2.1
Spain	13,132	5.8	1.8
Portugal	12,015	5.7	2.0
Greece	10,950	6.2	1.5
Hong Kong	21,201	5.5	5.1
Singapore	20,983	4.3	6.1
Japan	19,582	8.0	2.5
Korea	12,874	5.2	6.8
Malaysia	7,764	2.8	4.0
China	4,551	2.9	6.0
Philippines	2,369	1.8	0.8

Sources: Crafts (1998) updating Maddison (1995).
[a]Converted to $ using 1990 PPPs.

much higher in Europe than in the US but the UK proved to be an exception to this. From 1973 the European growth advantage over the US had largely been eliminated despite the fact that at the end of the period the US still maintained a large advantage in GDP levels.

A great deal of research in economics has been devoted to the study of the causes of differences in levels of income and levels of income growth. The availability of factors of production, their relative quality and the efficiency with which they are used are all partial explanations of these differences, as are disparities in the level and growth of codified and tacit knowledge in the economy. Useful knowledge, at least in terms of the determination of income levels and growth rates, can be produced by research funded by the state, investigations instigated by firms and by continuous improvements in ways of doing things that evolve from the learning associated with productive activity. This book attempts to address all these issues and aims to give a careful and thorough overview of the state of research into the determinants of productivity levels and growth rates. In particular we address issues of measurement and the role of knowledge.

The measurement of productivity and economic performance is well known to pose difficult and in some cases intractable problems. In some sectors, namely agriculture, mining, manufacturing, utilities, transport and communications, the volume of output appears to be relatively easy to measure, but such sectors have been becoming less important over the last seventy years. These sectors have shrunk from about 50 per cent of total US production in 1947 to about 30 per cent in the mid-1990s.[2] The sectors where measurement is particularly poor, i.e. education, health and government services, now accounts for about 30 per cent of US output. Before we can answer questions on productivity and performance we must first address the issues involved in measurement and comparability. Much economic growth has been driven by the accumulation of implicit and explicit knowledge, some embodied in individuals or equipment as factors of production, some available at a more abstract level, reflected in the scientific and technical knowledge base. This has increased the number of goods and transformed the nature of others, compounding measurement problems.

The papers in this volume are specifically concerned with explaining aspects of growth performance in the postwar world. Some of the papers, such as the contributions by Stephen Broadberry, by Geoff Mason, Brent Keltner and Karin Wagner and by Gavin Cameron, James Proudman and Stephen Redding are international in that they present evidence using data for more than one country, whereas many of the remaining papers are concerned with developments within one particular country. Our concern is with the determination of productivity and incomes in industrial countries, in part because detailed statistical analysis requires access to data of sufficient quality that are only available for a limited number of countries. But there is sufficient variation in the performance of these countries to provide us with interesting problems in their own right.

Examination of GDP per capita, while useful as a measure of well-being, does not convey enough information on cross-country differences in productivity, which we would define to mean output per unit of productive input(s). But once we move away from GDP per capita we encounter new problems involving how to define productivity and how to measure it. Thus many research papers maintain an interest in labour productivity – output per unit of labour input – reflecting the close links between this measure of performance and real wages and living standards in different parts of the world. Productivity can also be defined to reflect output per unit of standardised and aggregated inputs of all factors of production. A third definition takes productivity as the residual when account is taken of changes over time or differences across countries in both the quantity and quality of factor inputs. We will discuss both sets of approaches in this introductory chapter.

A number of measurement problems exist even in the study of labour productivity. On the labour input side, there are problems in the measurement of persons engaged (the treatment of multiple-job holders, the extent to which unpaid family workers are included in official figures) and more importantly in hours worked. From an economic point of view, it may be argued that labour input is more accurately measured as the time spent in productive activities. This would imply that, in evaluating trends over time or in carrying out international comparisons, we should allow for differences in the proportion of the workforce who choose to work fewer than 'normal' hours. In addition we should allow for time devoted to leisure activities such as holidays and the extent to which workers' productive time is lost due to sickness, maternity leave, strikes etc. The standard working week also needs to be adjusted to include overtime and short-time hours. Thus a great deal of information is required to measure the number of hours worked each year.

Similarly many problems beset the measurement of real output. Should output be measured gross or net of material input? The advantage of the 'net' or 'value-added' approach is that output can be summed across sectors but at the cost of quite stringent assumptions on the form of the production function. There are also particular problems in measuring output in large sectors of the economy such as government services, health and education. It has been common practice among statistical agencies to measure output in some or all of these activities by their inputs, and hence it is not possible to define productivity growth. If service delivery becomes more efficient this method of measurement will not reflect it, nor will the level of aggregate productivity. Furthermore, the extent of this practice has varied both across countries and over time, vitiating many of the comparisons we would commonly like to make.

Despite these measurement problems we do have some fairly reliable estimates of labour productivity performance across some industrialised countries, as shown in table 1.2. If we first consider differences in GDP per hour for the total economy, there are interesting comparisons to make with the data presented in table 1.1. In 1996, both

Germany and France had labour productivity levels slightly above those in the US in the total economy. This represents a major improvement from the poor relative position of these two countries in the immediate postwar period. Thus moving from GDP per capita to GDP per hour worked eliminates the apparent US advantage. This in turn is due to a number of factors such as lower unemployment, higher labour force participation and longer working hours in the US. But this is not to say that these factors as such explain the US lead in GDP per capita. These differences occur because of underlying disparities in the market environments for both outputs and inputs including both the degree of competition and the regulatory environments facing agents in different countries.

The third large European economy shown in table 1.2, the UK, has not, however, enjoyed similar gains in labour productivity terms over the postwar period. It had less scope for catching up because its labour productivity in the early 1950s was closer to that in the US than the other three countries shown in table 1.2. But productivity in France and Germany overtook that in the UK before 1973 and they continued to lead that country in terms of growth rates, albeit at a reduced rate, thereafter. The explanatory factors behind Britain's failure to keep up with her European competitors are again complex and are considered in some of the contributions in this volume.

In table 1.1 Japan enjoyed levels of GDP per capita of about 83 per cent of US levels in 1996. In contrast average labour productivity in Japan was only 74 per cent of US levels, even less than in Britain. Again this can be accounted for by labour market factors such as considerably longer working hours in Japan. Since Japan started the period with productivity levels only one fifth as large as in the US, it is not surprising that Japan also experienced the highest labour productivity growth rates, but the extent of Japan's growth lead has diminished since 1973.

All countries show a reduction in labour productivity growth post-1973, although the decline was not very marked in Britain. There are also some differences in the pattern of labour productivity performance in the market versus the aggregate economy.[3] In all five countries labour productivity growth in the market economy was nearly half a percentage point per annum higher over the entire postwar period than in the total economy. At first sight this may seem like a minor difference. But if this were purely a measurement effect, and productivity growth in non-market services was in reality no less than average for the remainder of the economy, then real average living standards in

Table 1.2 *Labour productivity[a] performance in the aggregate and market economies*

	US	UK	France	Germany[b]	Japan[c]
A. Total economy					
Labour productivity growth (% per annum)					
1950–96	1.55	2.60	3.70	3.87	4.48
1950–73	2.34	2.99	4.62	5.18	6.11
1973–96	0.77	2.22	2.78	2.56	3.06
Labour productivity levels (UK = 100)					
1950	100	51	41	37	20
1960	100	51	48	49	24
1973	100	59	69	71	44
1979	100	65	84	85	51
1989	100	75	103	84	65
1996	100	83	109	107	74
B. Market sectors[d]					
Labour productivity growth (% per annum)					
1950–96	1.97	2.97	4.08	4.07	5.02
1950–73	2.74	3.38	5.25	5.44	7.20
1973–96	1.20	2.55	2.91	2.70	3.13
Labour productivity levels (UK = 100)					
1950	100	49	35	39	17
1960	100	48	43	49	21
1973	100	57	63	72	41
1979	100	63	77	88	46
1989	100	73	95	96	58
1996	100	78	94	102	63

Source: derived from data in O'Mahony (1999).
Notes:
[a] Value added per hour worked.
[b] Former West Germany.
[c] The data series for Japan start in 1953.
[d] Excluding health, education and government services.

1996 in all countries would be about 15 to 20 per cent higher than measured average living standards. Cross-country comparisons are also affected by the move to a market economy basis, France now falls marginally behind the US whereas Germany's lead remains, but it is very small. The relative position of Japan falls further behind other countries when assessed on its market economy.

In conclusion, this brief look at labour productivity performance in industrialised countries shows that there are significant differences across time and countries which cannot be fully explained by a simple convergence model. In addition, focusing on differences between the aggregate and market economies suggests that measurement issues can have large effects on our perception of productivity changes and differences across countries. It is perhaps the case that the market sector has fewer measurement problems than the excluded sectors, and hence we should perhaps put more weight on these figures. However, even then problems remain as we have only considered productivity per unit of labour utilised.

When productivity is defined as output per unit of two or more factor inputs the measurement problems multiply as a number of measurement and definitional issues need to be considered. In the case of physical tangible capital assets we have to decide whether these are best measured as gross or net capital stocks or as a flow of capital services. (And then, how do we convert investment flows into measures of capital stocks?) How do we treat investments in intangible assets such as research and development (R&D) expenditures, advertising, marketing, and so on? To what extent can we take account of quality changes in both inputs and outputs? Finally there is a need to combine inputs in order to evaluate their impact on productivity and growth, that is, choices have to be made about which set of weights to attach to various inputs. Should inputs be weighted by the private returns they receive on the market or should we make some allowance for spillover effects or complementarities between inputs?

The next section presents a brief review of the literature to date on modelling and measuring productivity growth, thus placing the papers in this volume in an historical context. These papers are concerned with productivity and growth only in advanced countries and cover a wide range of topics including productivity measurement, diffusion of innovations, social capability and spillover effects. We consider the main contributions of the papers under a set of headings, but many of the papers cross the boundaries between sections and are included in more than one. We begin by looking at the measurement of inputs and outputs with a particular focus on issues to do with service sector productivity, and we touch on the role (or lack of it) for information technology in the production process. Much of the increase in output we have seen in the last two centuries or more has been driven by improvement in the quality of labour caused by the accumulation of knowledge, largely through formal education, but also through training whilst in the workplace. We then go on to look at the role of human capital in the determination of productivity, and at the associated impact of educational structures and the scientific research base.

The government has played a major role in encouraging and financing education and scientific research, and social capabilities must be seen as at least in part the result of government actions. However, technical advances do not come only from the traditional university and research institute based scientific community. Firms undertake a considerable amount of R&D, and the knowledge they create increases output and productivity. We discuss these issues, and the determinants of private sector innovations, before discussing the diffusion of ideas and the spread of technical change. Once knowledge is constructed it is essentially a free good, and it is welfare enhancing if it spreads quickly. Hence diffusion processes have to be understood. However, if information, once created, is freely available then there is no private sector incentive to invest in discovery. Hence there is a delicate balance to be struck between rapid diffusion and rapid innovation, and we discuss this policy problem for governments interested in growth and welfare.

Research on productivity: background

Although research in this area ultimately dates back to Adam Smith and other classical economists, modern approaches to aggregate productivity studies were heavily shaped by Robert Solow's seminal articles in the 1950s. These set out the neoclassical production function framework on which most subsequent work has been based (Solow, 1956) and elaborated the theoretical basis for the decomposition of output growth into a proportion attributable to measured increases in factor inputs and a residual, termed 'technical change' (Solow, 1957). Output was seen as determined by the quantities of factors of production available and by the level of knowledge in the economy. Countries differed in their output levels either because they did not have the same level of capital per unit of labour or because their knowledge stock was limited. If countries displayed the same saving rate, and hence the same level of investment, then convergence of their levels of incomes depended on the time it would take to reach the steady state capital output ratio and the rate at which they absorbed the knowledge in areas where they lacked it. Much of the subsequent aggregate productivity agenda can be seen as attempts to refine or criticise this simple description, and in the process useful insights from more disaggregated studies improved understanding significantly.

Beginning in the 1960s, in tandem with theoretical refinements of the Solow growth model, a strand of literature developed which sought to refine the Solow decomposition of growth into separate effects deriving from changes in the quality as well as quantity of production inputs.

One strand of this 'growth accounting' research was concerned primarily with the practical measurement of these decompositions, and this is best reflected in the seminal works of Denison (1967) and Maddison (1982). Considerable efforts were devoted to devising functional forms under which index numbers of productivity could be constructed which satisfied a number of theoretical properties, and this is best represented in the research on superlative index numbers by Diewert (1976) and others. The work of Zvi Griliches and Dale Jorgenson considered many of the details of weighting factor inputs, in particular the construction of indices of capital services, as in Jorgenson and Griliches (1967) and numerous papers by Jorgenson and his collaborators such as Dougherty and Jorgenson (1997). Many of the debates during this period centred on the extent to which the large absolute size of the Solow residual – as originally measured – reflected inadequate measurement of the changing quality of factor inputs. Subsequently, the quality-related adjustments to US input data series proposed by Jorgenson and his collaborators did indeed have the effect of sharply reducing the size of the estimated residual; much of this work for the US is represented in Jorgenson, Gollop and Fraumeni (1987).

Growth accounting is of course based on standard neoclassical assumptions that all markets are competitive and that all factors in the production process are paid their marginal products, the sum of which exhausts all returns from pursuing those activities. Other authors have concentrated on adapting the specification to situations where the original neoclassical assumptions did not hold, such as the incorporation of non-constant returns to scale, markups due to non-competitive behaviour and refinements to include the impact of regulation (see for example papers in Cowing and Stevenson, 1981).

Much of the early work on growth accounting was concerned with measurement of productivity growth rates over time in individual countries. At the same time there was a considerable body of research being carried out, largely by economic historians, on how innovations and technologies are diffused across nations. Many were in response to the entrepreneurial failure hypothesis expounded in Landes (1969) (see for instance chapters in McCloskey, 1971, and in Jeremy, 1992). An important theoretical treatment of the international diffusion of technology in history is David (1975). Furthermore, in the 1980s following the construction of historical time series by Maddison (1982), a substantial literature developed which focused on convergence of productivity levels across countries.

As is clear from our discussion, one implication of the Solow growth model in its simplest form is that levels of income per capita in all countries would converge and that long-run steady state growth rates would be constant across countries. Using Maddison's historical dataset, Baumol (1986) showed that there had in fact been convergence to the productivity leader, the United States, during much of the post-Second World War period among currently industrialised states. This perspective served as a substantial corrective to those commentators, largely in the media, who had predicted dire consequences for the US economy because of its relatively low growth in productivity by international standards.

The development of the convergence perspective precipitated much new econometric work in the area of cross-country comparisons of productivity levels and determinants. Much of this empirical literature exploited the Penn World Tables whose construction is explained in Summers and Heston (1991). Many of the initial papers were concerned with seeing if the neoclassical growth model, augmented to include inputs such as human capital, were adequate descriptions of the growth process, for instance Levine and Renelt (1992), Mankiw, Romer and Weil (1992) and Islam (1995) to name but a few, and in determining the speed of convergence as in Barro and Sala-i-Martin (1995). Much of this empirical work emphasises differences between advanced industrial and developing nations and subsequent research attempted to explain these differences by systems of government and social infrastructure, as can be seen from reading Rodrik (1998).

Convergence, and the role of institutions in it, is also of importance when discussing European growth experience, as the papers in Crafts and Toniolo (1996) make clear. The accumulation of factors and the acquisition of knowledge have been the dominating forces behind European growth, and national systems of education and knowledge acquisition have been central in aiding growth in countries such as France and Germany. Olsen (1996) in the same volume extends his thesis that institutions can retard growth, and that the development of strong and embedded interest groups prevents the restructuring of the economy that continuous innovation requires. In particular Olsen argues that the destruction of institutions in Europe (but not the UK) as a result of the Second World War meant that blocking coalitions of interest groups were removed and growth prospects enhanced. By contrast, Paque (1996), also in the same volume, argues it is too easy to assume that institutions changed dramatically after 1945, and it is probably wrong to ignore the creation of social coalitions that are designed to enhance growth.

Institutions, organisation and the absorption of technology produced by others can clearly aid the growth process. But as argued by Dougherty and Jorgenson (1997), relative productivity performance among advanced countries largely appears to reflect differences in capital investment of different kinds, both tangible and intangible. They argue that there is little scope for growth without the application of effort and the enduring of sacrifices. Inter-country differences in physical, human and R&D capital formation may in turn reflect established differences in incentive structures and institutional factors which condition investment decisions by enterprises.

The interest in convergence as an engine of growth also had the consequence that the calculation of relative productivity levels, not just growth rates, was put at the heart of productivity measurement. Research into measuring relative productivity levels also had a long history dating from Rostas's studies at the National Institute of Economic and Social Research (NIESR) in London in the 1940s (Rostas, 1948), to Paige and Bombach's (1959) work at the OEEC, the precursor to the OECD. The tradition has been carried on both at the University of Groningen, the Netherlands, under the tutelage of Angus Maddison and has continued to be at the heart of work on productivity at NIESR. Some of the more important issues are addressed in Prais (1995) and recent estimates across five countries can be found in O'Mahony (1999).

The theoretical study of economic growth was revitalised at the end of the 1980s with the first round of endogenous growth models which concentrated on the magnitude of the impact of capital accumulation, in particular clarifying the role of human capital (e.g. Romer, 1986; Lucas, 1988; Barro and Sala-i-Martin, 1995). Subsequently theoretical modelling moved more in the direction of understanding the processes of endogenous technical change (e.g. Romer, 1990; Grossman and Helpman, 1991; Aghion and Howitt, 1992). More recently models have moved to incorporating complementarities between inputs and technical change as in Howitt and Aghion (1998). These models have been exhaustively summarised in previous surveys of the literature such as Barro and Sala-i-Martin (1995), Grossman and Helpman (1991), and Aghion and Howitt (1998). However, the new literature has once again brought the study of technology and knowledge back into the mainstream of aggregative economics. Early attempts to look for increasing returns to scale have been replaced by a concentration on Schumpeterian forces of creative destruction and the creation of knowledge by maximising individuals rather than the altruistic state. Processes of diffusion and the existence of external economies of scale from knowledge have also re-entered the mainstream. The 'New Economic Geography' has emphasised processes of cumulative causation and studied the possibilities of irreversibilities in the growth process. These irreversibilities can come from externalities in the growth process, agglomeration effects and the effects of proximity in the diffusion of knowledge. Many of the ideas have been formalised by Fujita, Krugman and Venables (1999), relying heavily on developments in the theory of imperfect competition and its implications for trade that stemmed from the seminal paper by Dixit and Stiglitz (1977).

The Schumpeterian tradition in economics has been alive, and making a contribution to our understanding of the process of creative destruction, for decades. The rigorous mathematical modelling of the process of change is a fairly recent addition to 'mainstream' economics, but many of the ideas have been the subject of extensive research in work under the general umbrella of the study of technology. The study of innovation was not seen as a search for an explanation of the Solow residual, but rather as a search for the factors that affected the creation of new technologies that were an obvious source of increases in the ability of firms and industries to create output. Much of this literature was firmly placed in microeconomics, as can be see from the survey in Dosi (1988). The study of firms and their behaviour was central to Freeman's (1974) survey of the economics of industrial innovation, and to the body of work linked to studies of business cycles driven by innovation, as in Freeman (1983) and Freeman and Soete (1987). This work has been based around researchers at the Science Policy Research Unit at the University of Sussex. They have been considering these ideas for many decades, and it has also been addressed in the work of evolutionary economists stemming from Nelson and Winter (1982) and many authors within economic history. Much of this work arises from a common base with that underlying recent endogenous growth models: Schumpeter's (1934) *Theory of Economic Development*. One of the benefits of endogenous growth theory is that these ideas have re-emerged in mainstream economics, enriching our understanding of the economies in which we are presently living.

The theoretical developments in growth theory pushed aggregative economists in the direction of investigating the effects of research, innovation and knowledge on the growth process. In tandem with these theoretical considerations arose an empirical literature whose aim was the econometric estimation of the impact of technical progress on productivity. Much of this literature, which stems from the study of agriculture in the 1940s and 1950s, developed independently of growth theory and was avowedly microeconomic in nature. For example, work by Zvi Griliches, recently summarised in Griliches (1998), included cooperation with aggregative economists such as Jorgenson, but much of its focus has been on the impact of innovations at the microeconomic level and on the behaviour of firms and their role in the innovation process. Although strongly related to the work by Freeman and others, it was much more avowedly neoclassical and econometric in its orientation. The papers collected in this volume are influenced by these traditions.

In this introduction we first discuss the role of measurement and the problems in defining outputs and inputs. Over the last few decades data have deteriorated in quality and our conceptual categories for analysing them have not kept pace with change in the economy. One example of this is the role of computers in the production process. Until the mid-1980s the volume of computers in use in the US was indexed on their weight, rather than their capacity. When they were properly accounted for in the national accounts in the late 1980s they were the only investment good where 'hedonic' indices, advocated in Griliches (1961), were used to measure quantities in relation to capacities. This change increased measured productivity in manufacturing, but failed to have the same effect in services, which represent a much larger share of output. This IT paradox could reflect the problems users have in adapting to new technologies, and hence their introduction may take a long time to come through, a view in line with evolutionary and Schumpeterian views of the world. Griliches (1994) also suggests that the IT productivity paradox could just reflect our failure to properly measure output in the 70 per cent of the economy covered by services. We would have to move to hedonic, or quality adjusted output in services before we could fully reconcile this paradox.

The existence of national systems of education and scientific research has led to specific patterns of technology growing up in different countries. It is not just through the sponsorship of research that the government has a key role to play in constructing national capabilities. The decision by the German state in the mid-nineteenth century to construct educational structures in science and engineering is well documented, for instance in Freeman (1996). These were consciously designed to ensure that Germany caught up with industrial productivity in the UK, and they were important in ensuring that it did. The formation of human capital is an obvious and central way of ensuring that scientific knowledge becomes embodied in the production process. The national science base is more important in some countries than others, and the role of public sponsorship differs between countries. R&D in the private sector is particularly strong in the US, reflecting both the greater role of private enterprise and the clearer nature of shareholder driven interests through the stock-market valuation of companies. The study of company data on innovations and their integration into more aggregative work remains a core part of any research agenda in the study of productivity.

The importance of the firm in the process of innovation is stressed by both the econometric study of the behaviour of firms and the construction of firm level databases on innovations such as that at SPRU. However, the study of national systems of innovation, of firms and of innovations, is not enough to complete our picture of the factors affecting growth. There are spillovers from R&D in one firm to output in others to consider, and the role of external effects from knowledge, which is also addressed in this volume, has been at the centre of research. It can be seen in work on firms such as that by Jaffe (1986), as well as in the work of neo-Schumpeterians, with Archibugi and Michie (1997) providing a compendium, and has been reinvented in the endogenous growth literature in the last decade or so. However, we have much to learn about the forces behind, and ports for, the diffusion of technology, and the role of the international firm and the relation of technology to trade are vital topics to investigate.

Service sector productivity: measurement and explanation

As discussed above, the difficulties associated with productivity measurement in manufacturing industries are multiplied many times over in service sector activities. They are particularly severe because of the problems relating to the measurement of output. It is hard to find physical measures of service outputs or to identify the appropriate price data needed to deflate sales or revenue measures of activity. However, even before considerations of data availability, it is often far from clear how the 'output' of a particular service should ideally be defined. For instance, should the output of the retail industry be valued primarily in terms of the number of transactions or the total value of transactions? Should the services provided by doctors be appraised in terms of numbers of patient-visits or the numbers of patients cured? Value added indicators in constant or current prices are deficient in this respect. The former requires solving the problem of the price of a unit of output whilst the latter would require the determination of plain value added, and where there is no market, such as in the provision of public services, it is conceptually difficult to do this.

Even in those service industries where the conceptualisation and measurement of output appears to be straightforward, the nature of many service activities is such that improvements in their quality as experienced by customers are often associated with increased labour-intensity and thus in a reduction of (crudely) measured labour productivity. For example, upgrading the quality of services in hotels may take the form of employing more staff to carry customer luggage and to offer room service over a longer period of time each day. Similarly, an obvious way of improving the quality of service provided by telephone help lines is to employ more people in order to reduce customer waiting times.

Nevertheless the existing data in official statistics do convey some information on productivity performance in service sectors. The second chapter in this book, by Stephen Broadberry, highlights some important trends. He suggests that the well known deterioration of Britain's aggregate labour productivity performance relative to countries like the United States and Germany over the past century may owe far more to changes in relative performance in services than, as widely believed, to any long-term relative decline in British manufacturing productivity. Broadberry sets the historical background by presenting estimates of relative sectoral productivity in the US, the UK and Germany over the period 1870 to 1990 which have been constructed from official and historical national accounts sources. In terms of aggregate labour productivity, Britain was overtaken by the US towards the end of the nineteenth century and by Germany in the late 1960s. However, in manufacturing the US productivity lead over Britain in the late 1980s was much the same as in the 1870s. German productivity levels in manufacturing were broadly similar to those of Britain in the first few decades of the twentieth century. Subsequently, Germany overtook Britain in the 1950s and thereafter, apart from a brief but significant German advantage in the 1970s, the manufacturing productivity gap between these countries has remained relatively small.

Britain's relative decline in aggregate productivity performance appears to reflect two core forces. Firstly, the fact that shifts of labour out of agriculture occurred much later in the US and Germany than in Britain; and secondly, the gradual erosion of the lead which Britain enjoyed in service sector productivity in the late nineteenth century. As described by Broadberry, the main factors underpinning Britain's loss of its once sizeable productivity advantage over the US and Germany in service industries included:

- the gradual convergence of urbanisation levels – and associated economies of agglomeration – in the three countries;

- the US and Germany also both benefiting from development of commercial organisations based on hierarchy which outweighed Britain's early advantages from 'network' forms of organisation; in addition there were significant developments of US and German advantages in different forms of human capital formation;

- the fact that Britain was 'locked in' to slow-growing imperial markets after the Second World War and also other 'lock-in' effects due to Britain's relatively early accumulation of physical capital. These had the effect of retarding productivity growth in service industries such as railways.

In Chapter 5, Mason, Keltner and Wagner tackle some of the measurement difficulties associated with service sector productivity through a detailed comparative study of commercial bank lending in three different countries: Britain, the US and Germany. Their findings are based on data collected from matched samples of bank offices engaged in lending to 'middle market' or 'mid-corporate' business customers. After standardising for average loan size, average lending output per employee-hour in the German sample is estimated to be some 23 per cent higher than in the US and almost two thirds higher than in Britain. This productivity ranking is left undisturbed by consideration of various indicators of service quality as experienced by borrowers, such as banks' speed of response to customers' credit requests and the extent to which banks seek to gain a detailed understanding of clients' businesses and their associated credit needs.

Mason, Keltner and Wagner's focus on a narrowly-defined service activity coupled with primary data collection at an establishment level enables detailed consideration of the impact of certain conceptual and definitional choices which have to be made in the course of service productivity measurement. For example, their central estimates are based on a definition of loan output in terms of completed loans rather than the number of proposals that were formally prepared by lending staff. The labour inputs associated with lending are defined to include time spent in efforts to develop new customer contacts as well as time devoted to servicing existing contacts. In this context the measured productivity performance of the US sample of lending offices in particular is diminished by a higher proportion of time lost in preparing formal proposals which are not accepted by the customers for whom they were intended. This rejection by customers appears to reflect a higher level of price competition from other lending institutions in the US than occurs in Britain or Germany. Measured productivity performance is also diminished by the larger average proportion of time which American relationship managers devote to searching for prospective new clients.

These findings shed light on the difficulties in interpreting inter-country comparisons of banking productivity based on published data sets which cannot distinguish between labour inputs directly devoted to the production of bank service outputs and other labour inputs which are absorbed in responding to competitive market pressures. Business strategies adopted by US mid-corporate lending offices apparently contribute to higher average net income per employee-hour in the US than in either of the two European countries. However, they are also associated with greater expenditures of American lending managers' time on activities which are essentially 'unproductive' in terms of their contribution to measured bank output.

Improved productivity in services impacts on the productivity of other sectors that buy services, and the pass-through of new technologies cannot be ignored. In Chapter 3, Greenhalgh and Gregory emphasise the importance of business services in the process of diffusion of new technologies throughout the economy. This sector is important because it provides an intermediate input for other sectors rather than a final consumption good, and in certain circumstances its output would be consolidated into manufacturing, as firms have the choice of buying in such services or providing them directly in-house. The main features of this paper are considered further below.

Measurement issues are also at the heart of recent debates concerning the apparently slow growth of productivity in financial services and other service industries which have been among the heaviest investors in new IT (information technology) equipment in recent years. These issues are addressed in a very direct way in Chapter 4, by Michelle Haynes and Steve Thompson, which examines the impact of a single homogeneous embodied IT application (ATMs – automatic teller machines) on labour demand and productivity in the UK building society sector. Building societies, equivalent to Savings & Loan institutions in the US, are seen as especially suitable for avoiding the problems created by multiple outputs because they were limited to core activities of deposit collection and mortgage finance until 1987. Thereafter, firstly, diversification was limited and, secondly, separate accounts were maintained for core activities.

Haynes and Thompson start with a careful discussion of the conceptual and practical difficulties of separating out the effects of IT investments on productivity from other determining factors such as the quality of management and changes in work organisation. They then present estimates of labour demand equations in which ATMs are treated as a completely new and separable capital input. After controlling for output, wages and lagged employment, the ATM terms are found to enter with negative signs and to be statistically significant, and hence they can be described as a source of technical progress as they are reducing labour inputs needed to produce a given level of outputs. These employment effects are even greater for intensive ATM users than for non-intensive users. Indeed, the falls in employment associated with intensive ATM adoption are so great that the authors suggest this process innovation may be correlated with other unobservable IT inputs. At the same time ATMs have *product* innovation characteristics which may have helped intensive adopters to achieve exceptionally fast growth in output and in market share during the 1980s market upswing.

Haynes and Thompson conclude that their findings give no support to common views of an 'IT productivity paradox'. However, for various reasons, they suggest that their specific results may be regarded as upper bound estimates of the employment-reducing effects of IT investments in the UK building society sector. They also suggest that early adopters of ATMs may have had more to gain from adoption than other potential users and that improved service quality may have led to early adopters gaining market share at the expense of non-adopters. Thus not all productivity gains achieved by early ATM adopters will necessarily be available to later adopters.

Human capital formation, educational structures and the scientific research base

In their explanations of why less-developed countries have failed to participate alongside advanced industrialised nations in the process of convergence on the productivity levels of the United States, researchers now typically allocate a key role to inter-country differences in education and in skill endowments. For example, technical, organisational and administrative competencies are important ingredients in the notion of 'social capability', which may differ systematically between nations with pronounced effects on their ability to acquire and make use of new technology (Abramovitz, 1989). Inter-country differences in the human capital assets required to benefit from cross-country spillovers of innovations and investments also play a central role in endogenous growth models of capital accumulation, as surveyed in Barro and Sala-i-Martin (1995).

A number of recent papers have highlighted the importance of human capital in the development of social capability. Skill levels affect the ability to innovate and to adopt technology from abroad, and hence affect the growth rate as well as the level of productivity (Benhabib and Spiegel, 1994). Much of this work is focused on comparing the social capability of advanced countries with that of countries at lower stages of development, although the insights are important in understanding long-run differences between advanced economies as well. Empirical studies of the educational components of social capability suggest that a high average level of attainment at the secondary level of education is an important pre-condition for rapid growth. The studies are usually broad in their coverage, with many developing countries included in their samples, and tend to find little role for attainments at primary level or indeed at higher education level (Baumol, Blackman and Wolff, 1989; Barro, 1991).

However, evidence has accumulated that, even among advanced economies, the observed pattern of difference in national economic performance may in part be attributable to cross-country differences in human capital endowments, educational structures and the scientific research base as well as other aspects of social capability. For example, several researchers have noted the ways in which US manufacturing productivity leadership has been assisted by the ready availability of well-qualified engineers and future managers emerging from the long-established American system of mass higher education (Chandler, 1992; Broadberry, 1994; Mason and Finegold, 1997).

Others have highlighted the distinctively close relationships between many US enterprises and universities. For example, leading American universities were far in advance of their British and German counterparts in developing separate disciplines of chemical and electrical engineering designed to meet the needs of large-scale US producers in those fields (Landau and Rosenberg, 1992; Nelson and Wright, 1992). Some, for instance Stanford with its large foundation, have even been prepared to play a more proactive role and provide venture capital for science based small firm setups. At the same time, within Western Europe, cross-country comparisons of matched samples of establishments carried out by researchers at the National Institute of Economic and Social Research (NIESR) in London have identified strong national differences. There are various mechanisms by which, for example, the relatively high levels of craft skills generated by the deeply-rooted German apprentice training system contribute to superior productivity performance in German manufacturing plants relative to counterpart British establishments (Prais, 1988, 1995).

Research interest in the differences in national education and training institutions between advanced countries, and the economic impact of such differences, has arguably intensified as evidence has mounted of renewed growth in the individual returns to skill and knowledge acquisition within advanced economies (Katz and Murphy, 1992, for the US and Machin, 1998, for the UK). Frequently this literature suggests that technical change has become more skill biased over time. Much of this research points to the existence of important complementarities between high-level skills and the diffusion of advanced technologies, in particular, microelectronics-based equipment (see Caselli, 1999, for a recent overview of US and British evidence in this area).

An earlier literature centred on the ways in which education may enhance the economic value of individuals. For example, it may work by increasing their ability to respond to new opportunities and 'deal with disequilibria' (Schultz, 1961, 1975). Alternatively it may be that educational qualifications operate as screening devices for the personal attributes sought by employers (Spence, 1973) and their efficiency does not decline as higher education expands.

Researchers investigating the sources of differences in productivity and growth performance between different Western countries need to be able to take full account of national-institutional differences in the development of workforce skills and knowledge. These are considered to some extent in the chapters by Broadberry and Mason *et al.* In

Broadberry's chapter the development of mass higher education in the US and highly organised apprenticeship training in Germany in the twentieth century is seen as crucial in overcoming an earlier UK advantage in the extent of professional qualifications, in particular in service sectors. Mason *et al.* show that in banking a high proportion of German clerical employees have undergone a full apprentice training after leaving school (in contrast to their generally less well-trained American and British counterparts). The German banks had the lowest proportion of clerical/secretarial staff in their samples. Chapter 6, by Cameron, Proudman and Redding, discussed further below, also finds that human capital is important in facilitating the process of technology transfer.

As our understanding of the growth process has improved, it has become clear that it is possible to get trapped into situations where cumulative causation leads to a lower level of output than might otherwise have been obtainable. Finegold and Soskice (1988) suggested that the UK may have become caught in a 'low-skills' trap, in that a low level of skills meant investment and research and development were held back by the lack of skilled labour. In turn, the decision to invest in skills, based on the earnings available, may have been adversely affected by low marginal productivity of labour driven by low levels of investment and R&D. This story depends crucially on the existence of complementarities between capital and skilled labour, and also possibly between R&D stocks and skilled labour.

Stephen Nickell and Daphne Nicolitsas in Chapter 11 attempt to shed light on this topic. They explore the hypothesis that investments in different forms of capital (machinery, workforce skills and R&D) are strategic complements of each other through panel data analysis of a large number of UK manufacturing companies. Their results suggest that skilled labour shortages may have cumulative and permanent effects on output via their effects on investments in physical capital. Nickell and Nicolitsas calculate that a permanent 10 percentage point rise in the reported incidence of skill shortages at the industry level would induce a permanent 10 percentage point reduction in physical capital investment and hence reduce the level of future output. Although skill shortages do not appear to have a permanent effect on investments in knowledge capital (R&D), they appear to contribute to temporary reductions in R&D expenditure via the dependence of R&D spending on output growth.

They look at two types of investment – in physical capital and in knowledge capital (R&D). They require very different approaches, in part because the latter form of investment is produced by stocks of very skilled labour, and firms tend to keep these stocks in place through downturns. As a result, R&D investment is a lot less cyclical than physical investment. In order to reflect these differences, the model they use for physical investment is heavily based on the theory of the maximising firm, while that for R&D is more eclectic, but both imply slow adjustment to an equilibrium.[4]

It is important to understand the framework within which R&D takes place, and there is a very good case for arguing that it is still organised on a national basis. The boundaries of the state and the nationality of firms are strongly related, and to date there has been a strong national bias in links between business enterprises and publicly funded research institutes. Both patenting and R&D expenditure data show that firms continue to perform a high percentage of their innovative activities in their home country. These biases give some guidance in the construction of policies for a 'National System of Innovation', which Pari Patel and Keith Pavitt in Chapter 9 define as national investments in knowledge-generating activities that are necessary complements to investments in equipment in increasing efficiency and in maintaining or increasing competitiveness. Such systems can benefit from public subsidy through the education system and through public sector research institutions. However, the best policies to adopt will vary with the size of the nation and the relative size of firms in its dominant industries, as large multinationals are more likely to undertake R&D abroad.

Much human capital is produced as part of the structure of education and research funded by the government in universities and research institutes. These form part of what can be described as a national system of innovation, and need to link very strongly to innovating firms and private sector research bodies. Patel and Pavitt argue that national systems of innovation are now coming under increasing strain from mismatches between skills and knowledge required by domestically-owned innovating enterprises and those provided by national science base institutions. This applies particularly to Germany and Japan. The UK has experienced relatively weak demand from domestically-owned enterprises for the services of universities and other national science base institutions. However, what otherwise might have constituted a major problem for domestic investments in high-level skills formation and knowledge transfer capacity has been offset to a considerable extent by an increase in links between the British science base and foreign owned firms.

Sources of technical progress: R&D and innovation

The general level of productivity in the economy is usually seen as depending on information and knowledge, and there are many sources for such effects in addition to educational systems and publicly funded research institutions. Individual firms and entrepreneurs are continuously creating new knowledge with the explicit intention of raising their profitability and probably therefore also the level of output. This may be reflected in new products or new equipment for producing their products. There may also be new processes, new ways of doing or organising things that improve the use of the existing factors of production or change the quality of the product. Ideas gaps and object gaps (to use the terminology in Romer, 1993) can be filled by product and process innovations. Technological leaders commonly produce innovations and they then spread, or diffuse, to followers. In this section we discuss the production of knowledge through conscious action, whilst in the next we look at the spread of knowledge through markets and market failure. In order to understand this process we need to look at empirical investigations of the factors affecting firms and their decisions. These factors have to be seen as an additional layer of explanation to that given by our understanding of the scientific research base and the national system of innovation.

Firms and entrepreneurs are continually investing in knowledge, as is the state, and much of this search can be described as expenditure on research and development. The incentive to innovate must in part depend on the ability of the individual to reliably expropriate the value of the innovation they have produced. However, the more of the rent on an innovation that is kept by the innovator the less the spillover to the rest of society. Knowledge, once created, is essentially a free good, because its use is non-rivalrous, but if all knowledge were immediately and freely available then the incentive to produce it would be dramatically reduced. There are several solutions to this problem, and as Schultz (1953) suggests, much of the important innovative activity in the US in the late nineteenth and early twentieth centuries was produced by the state through the network of agricultural research institutes it constructed. A great deal of general innovative activity is still undertaken by the state through universities and other publicly funded bodies, in part to deal with the public good, or market failure, aspects of basic scientific research.

However, the framework for innovation has been continually redesigned to recognise the need for private sector individuals to have incentives to innovate. Patent laws are designed to protect the innovator, but cannot be too restrictive or new ideas would not spread. The general level of scientific knowledge in the economy may affect the efficiency with which things are done without being embodied either in physical or in human capital. This can be from two sources, both of them external to the firm. We may see external economies of scale that raise the level of output available to the economy, and these will be reflected in the aggregate level of output. There may also be spillovers from the activities of one firm to another, and the level of R&D in an industry (or in the economy) may affect the output of all of the individual firms in the industry, even when some do not undertake any R&D, and these issues are discussed in Chapter 10 by Paul Geroski, Ian Small and Christopher Walters, who find large but uncertain effects from spillovers.

The evaluation of the impact of R&D on productivity is an essentially complex problem. Some of the impact of R&D comes through non-appropriable (and ostensibly non-measurable) spillovers to other firms, and the incentive to innovate may not reflect its social value. It is difficult to assess the impact of R&D spending because its effects come through only with a lag, and that is long and variable. Even patents, which may be more directly related to outputs, are a poor indicator because there are a large variety of reasons for taking them out. It is possible to define the Solow residual as the technical progress function, and look explicitly at the determinants of technical progress for the firm, the industry and for the economy as a whole. In order to do this we have to be clear what determines R&D and how it affects output.

However, if markets exist in which information can be evaluated, and if they work relatively efficiently, then the appropriable future value of current and past R&D expenditures and the resulting patents should be reflected in the market value of firms. If this is the case we may be able to assess the impact of such spending on productivity and on the incentive to innovate.[5] Bronwyn Hall in Chapter 7 surveys the literature on the effects on market valuation of R&D and patents, and reports on some new work in the area. She reports on the preliminary results of several projects, one of which finesses the problem of the relation between R&D spending and technological advance by using citation weighted patents as an indicator of the significance of a particular piece of knowledge. This innovation in data handling adds significantly to our understanding of the impact of R&D on the value of firms and on the productivity of the economy.

There are of course many difficulties in evaluating the effects of research on valuation, but it is probably as good a route to assessing their economic worth as any other. If an innovation produces a complex good then its elements need to be unbundled using hedonic indexes in order to assess the contribution of the research to its increased usefulness. The evaluation of R&D effects is perhaps directly measured through the use of hedonic techniques on the value of the company, although this will only incorporate the elements of R&D effects that can be appropriated. The value of a publicly traded company, at least where there is a thick and well informed market, as in the US or the UK, can be seen as a bundle of assets. If the market is well informed, and in equilibrium, then we can use regression techniques to evaluate the impact of inputs, such as the (depreciated) stock of R&D, on the value of the originating company. US estimates, surveyed in Hall, suggest that the stock of R&D is incorporated into the value of the company with a valuation ratio of between 0.5 and 2.0. This does not take account of spillovers, and the evidence Hall cites suggests that the absorption of other firms' R&D is aided by the existence of a large R&D department which can act as a port for managing and interpreting new information flows.

Earlier work by Hall suggested that the valuation ratio for R&D fell in the 1980s, potentially reducing the incentive to innovate and therefore adversely affecting the level of output in the economy in the long run. However, she finds that this fall in valuation was the consequence of events in a small number of industries, and that the overall valuation of R&D began to rise after 1991, albeit not back to the levels seen in the 1980s. There have been few studies on the UK, in part because of shortage of data, and those Hall surveys suggest that the impact is less than in the US. Few studies have used patents, and UK studies tend to use the SPRU innovations database, but in general the use of patents can be seen as adding to the use of R&D, although it is less directly important. Hall also presents preliminary findings from her recent work on the use of patent citations and suggests this is an important avenue of future research, It is clear that we need several indicators of scientific activity in order to assess its impact on the firm and its value.

There has of course been a gradual internationalisation of R&D in the postwar period, and hence the mainsprings of growth have become less nationally idiosyncratic. The combination of increasing pressures from competition that have resulted from the liberalising actions of domestic governments and reduction of barriers to trade have combined to force firms to think about appropriate locations for R&D activities. This process has been facilitated by the liberalisation of foreign trade and the reduction of barriers to the movement of capital. The growth of knowledge has also meant that the range of fields for useful technological knowledge has been increasing in the last five decades. As a result of these forces some researchers are in a position to argue that the internationalisation of R&D activities is increasing, in particular amongst advanced technology companies. Analysis in Chapter 9 by Patel and Pavitt of patent data for 220 companies with a high volume of patenting activities suggests that this process is only gradual in nature. Where it occurs it is most likely to take the form of firms locating technological activities abroad to support their strengths at home, perhaps as local adaptation facilities. Only occasionally do firms look to locate abroad in strong locations when they themselves are weak in the field in which the new location operates.

Both patenting and R&D data show that firms continue to perform a high percentage of their innovative activities at home. However, it is common for large firms from small countries to locate more of these activities abroad, and both UK and US firms have significant research bases abroad. Although there has been a modest increase in the percentage of R&D located abroad since the early 1980s, it cannot be seen as a major force in changing the distribution of incomes across countries. Developments in Swedish R&D are revealing, both for the causes of internationalisation and for the limits it faces. Braunerhjelm and Ekholm (1999) argue that Swedish manufacturing is unusually concentrated, with a small number of specialised firms. This has developed for historical and political reasons, and has resulted in a relatively large amount of foreign direct investment, and foreign production involved 52.2 per cent of output in 1994 and 61.1 per cent of employment. Inevitably a considerable amount of R&D activity would be located abroad in these circumstances. Foreign based R&D rose from 8.6 per cent of the total in 1970 to 24.7 per cent in 1994, whilst foreign production only doubled as a percentage of the total in this period.

The Swedish example is particularly revealing because, over the period considered, the number of graduates produced by Swedish universities stagnated, and at the end was actually falling. The combination of a compressed income distribution and some failings in the educational system meant that the stock of workers available to undertake research shrank in relative terms compared with other countries, and it is not surprising therefore that R&D had to begin to move abroad.

The gradual increase in foreign location of technological activities by large firms reflects the increasing range of potentially useful technologies that these firms need to master. Competitive pressure to search outside home countries for new knowledge is felt most strongly by large firms based in small countries. However, most large firms are aware that public science bases in different countries may vary in the effectiveness with which they get to grips with emergent technologies and this is a stimulus to worldwide knowledge search activities. In this context home-country R&D activity continues to be strong mainly because enterprises are mindful of the difficulties of managing development and commercialisation of major innovations across national boundaries.

The diffusion of technologies across companies, between sectors and among countries

Individuals may possess information that they wish to keep to themselves, and companies may have techniques whose rent they wish to guard, but ideas tend to diffuse to others. Knowledge spreads through scientific journals, through education – especially the internationalisation of graduate education over the last century – and through conscious decisions of governments and firms to relocate the structures that embed information. In most cases there are significant costs in the diffusion of innovations, and there have to be incentives to create knowledge and to move it. Trade is one obvious channel through which information about innovations flows. The purchase of new machines will embody new techniques, and the decision to penetrate a new export market will require the acquisition of knowledge about products and processes in that market, and hence will necessitate the acquisition of skills. Increasingly as production has become more science based, and more of the stock of R&D is the result of investment by firms, the diffusion of innovations has been through firms setting up new plants (or converting existing ones) in new locations to utilise their firm specific knowledge. This allows them access to new markets, seeking rents in this way. They choose not to do so through licensing as this can cause knowledge seepage. The diffusion of innovations by this route has been a major reason why the postwar world has seen the rise of the multinational firm to its current level of dominance, as is discussed in Barrell and Pain (1997). This book contains contributions that address the diffusion of innovations between firms, across industries and amongst nations.

The factors affecting productivity and growth should be reflected in the performance of individual companies, even where those factors reflect spillovers from other companies, sectors or countries. It should be possible to extract evidence on these factors from company level data sets. In particular, Geroski *et al.* argue that we should be able to discover whether performance depends on the actions of the firm or whether it is also the consequence of the environment in which it exists, with its productivity depending on the activities of other firms. There are two sets of factors that may make the macro outcome for a group of firms different from the outcome any one of them might produce. It is possible that there are external economies of scale, with firms benefiting from a common pool of labour or other resources that are dependent on the size of the industry rather than the size of the firm. If these come in the form of agglomeration economies, then they are often the result of the size of the industry, and we would expect to see strong correlations in productivity growth across firms in a sector or a location. It is also possible that the external economies could come from spillovers from innovating firms to the rest of the industry. In both cases the aggregation of data to the industry level is likely to increase the estimated degree of returns to scale. Increased output in one firm increases productivity in others, and hence output rises by more than can be expected from the effects of the increase in inputs on the output of the firm on its own. Hence aggregate studies may underestimate the degree of agglomeration economies and external spillovers in an industry. It is particularly difficult to measure technical spillovers, as the ability to undertake R&D spending is probably correlated with productivity, and as an input variable it may take some time to have an impact on firm productivity.

Geroski *et al.* address these aggregate problems by starting with a panel of 216 large firms over the period 1974 to 1990. They work with the Solow residual for individual firms and attempt to decompose it into the parts driven by systematic determinants and those that are idiosyncratic. Most of the variation in the residual comes from variations over time in company performance, and the variability within the (unbalanced) panel actually increases over time. The paper looks for the effects of industry-level productivity growth and economy-wide productivity growth in a dynamic framework. These are seen as determinants of firm-level productivity growth with the former additional variable representing specific agglomeration effects whilst the latter might reflect the effects of the scale of the economy on its ability to operate. Relatively large and robust effects from these variables are clear in the firm-level regressions, suggesting that there may well be agglomeration effects in the economy. However, the paper goes on to argue that these effects cannot be seen as robust to changes in specification. The authors look at various indicators of technical innovation and the capacity to absorb new information. Where data are available on innovations (from the SPRU database up until 1982) or on patents (up to 1988) it is used, but little effect is found. Spillovers from the activities of individual firms appear to be weak, although the effects of a generally more innovative environment cannot be captured in this framework. Various technical tests are undertaken of the robustness of the results, and the authors conclude that the initial estimates cannot be seen as strong support for external economies and spillover effects from technology.

Christine Greenhalgh and Mary Gregory (Chapter 3) also look at the diffusion of innovations through the production structure of the economy, and they find that services play a key role both as sources of labour productivity growth in the wider economy and as elements in its transmission. Using an empirical framework based on input–output tables for the UK, Greenhalgh and Gregory examine cost-savings and the transmission of R&D across sectors. Inter-industry transactions in intermediate goods allow linkages to be traced backwards from final use through supply chains and forwards from suppliers to purchases.

This disaggregated sectoral approach enables them to identify various ways in which increases in productivity in services contribute positively to productivity in other sectors. Changes in labour productivity are seen as rooted in the organisation of production and in the development of new products rather than in primary factor efficiency. Productivity growth is found to have been particularly high in business services and inputs from this sector have been increasing over time in all sectors. The latter partly reflects the increase in service inputs such as marketing, software development and increased use of information technology but also involves growth in contracting out of traditional services such as accountancy. These findings, which are unlikely to be confined to the UK, highlight the increasingly important contribution that service activities make to the general process of productivity improvement and growth.

Gavin Cameron, James Proudman and Stephen Redding (Chapter 6) assess the extent to which economic growth is related to the degree of international 'openness', defined as the magnitudes of impediments to the international flows of goods and services, factors of production and ideas. Drawing on the recent theoretical literature on endogenous growth, Cameron *et al.* suggest that international openness may have proximate effects on the rate of productivity growth through its impact on domestic rates of innovation. It also affects the amount of technological knowledge that can be imported from the leading economy in an area of production and the rate at which this technology transfer occurs. They find that different industry-level measures of openness (for example, ratios of exports, imports, foreign direct investments and trade weighted R&D stocks to aggregate activity) are positively correlated with both labour and total factor productivity but not with the capital–labour ratio. This suggests that openness affects growth through the rate of technological change rather than through investment. They then consider a theoretical framework in which a laggard country (the UK) has lower productivity levels than a country on the technological frontier (the US). In this model productivity can grow as a result of sector-specific innovation or as a result of technology transfer from the frontier economy. Their econometric results based on this model suggest that the rate at which a UK manufacturing sector's productivity converges to the US level is positively related to trade in goods and the spillover of ideas. At the same time an industry-level measure of human capital (the ratio of employees with high-level and intermediate-level qualifications to total employment) is found to be positively related to the rate of technology transfer.

Innovation and diffusion are strongly related, but not always fully understood. Chapter 8, by Paul Stoneman and Myung-Joong Kwon, explores the impact of technological change on gross investment, building a bridge between neoclassical investment models and diffusion models. The former commonly define a capital aggregate and then analyse changes in the demand for that aggregate over time whereas the latter first examines demand over time for individual capital goods and then aggregates. The diffusion approach allows the incorporation of a term in technological opportunities to enter the gross investment function. The paper presents models of diffusion at both the firm and industry level.

The firm-level diffusion model begins with the assumption that technological change requires the installation of new capital goods. Gross investment is made up of adoption investment, defined as investment undertaken to change technologies given the current level of output and expansion investment, defined as investment undertaken to expand output. Adoption investment is modelled using diffusion theory, and central to this is the assumption that the price of a capital good falls relative to the price of all capital goods the longer it has been available. The probability of a firm adopting a specific technology at time

t is related to both an epidemic effect whereby the longer a technology has been on the market the more likely it is known to the firm, and thus adopted, and to the net expected cost at time t of waiting until time $t + dt$ before acquiring the new technology. Hence Stoneman and Kwon derive an equation where real adoption investment is a function of the firm's level of output and a number of variables that impact upon the speed of diffusion including capital prices, expected changes in capital prices, firm characteristics and industry prices.

The number of technologies is proxied by a log linear function of firm R&D expenditures, firm patents and industry R&D expenditures. Firm characteristics include firm size, cash flow and the wage rate. Adding the expansion investment component, assumed to be a function of increases in output since the last peak, yields an equation for overall firm investment. The paper tests the model on data for a panel of firms in the UK. The results suggest that the R&D variable is significant and positive but patents are negative and insignificant. The results show that both technological opportunity variables and diffusion related variables impact significantly on investment. Estimates of long-run elasticities indicate that these variables are quantitatively significant.

The basis of the industry model is first to identify for industry k the threshold level of output at time t beyond which a firm in that industry will be a user of a specific technology. Reductions in this threshold level over time will trace out the diffusion path, which is the proportion of industry output produced using this technology, from which it is possible to derive gross investment in this technology. Summing over the technologies available at time t gives total gross investment in new technologies. This diffusion process yields an equation relating real adoption investment to industry output, the number of technologies and variables that effect changes in the average costs and benefits of adoption such as changes and expected changes in the price of capital goods and industry characteristics. In this model it is changes in the benefits of adoption, rather than the level of benefits in the firm model, that matter. The number of technologies are again proxied by a linear combination of a number of variables including industry R&D, the number of patents granted to firms in the industry, the number of trademarks newly registered to firms and numbers of innovations produced and used by firms in the industry. The data used is a panel of 23 UK industries from 1968 to 1990. Again Stoneman and Kwon find that technological opportunity impacts positively on gross investment. They conclude that the results imply that models of investment that do not include technological opportunities may be mis-specified. Of more importance, in the diffusion model technological change is central to the investment process rather than merely the result of adding a technological change term to a standard investment equation.

Conclusions

Our understanding of the forces driving technical progress and determining the level of productivity tell us that the knowledge stock is central to the process of growth. This stock is the global sum of ideas relating to basic and applied scientific work as well as non-scientific understanding of ways of doing and ways of organising. Its importance depends upon its application in both technical and non-technical activities. Technical change, or innovation, involves accessing the knowledge stock and using elements of it to embody them into an economic process. This volume addresses the role of knowledge in production and its diffusion within and between countries and sectors. There remains much to be done to improve our understanding of the role of knowledge in production. There are obvious gains to be made from further careful econometric work such as that by Haynes and Thompson, Geroski *et al.*, and Nickell and Nicolitsas on firm level data. However, careful case study projects, such as that discussed in Mason *et al.*, will also aid our understanding of the role of knowledge. It would be of great value to combine the two, and we see the merging of case study and econometric work as a major way forward in this area.

Knowledge is an interesting commodity, because once it exists it is non-rivalrous in use, and it is not diminished when used. However, it is clear that accessing knowledge is not a costless process, otherwise the speed of convergence of incomes across countries in the world would be much faster than we observe. Success in accessing and using knowledge is central to the promotion of growth, and there are clear advantages to leadership in this game. The costs of finding knowledge and the incentives to do so depend upon the infrastructure of the economy, on attitudes to risk both from individuals and by capital markets, and on the level and type of education in the society. Complementary inputs facilitate the absorption of knowledge, as do clusters of vertically or horizontally located firms with external economies from common pools of labour and knowledge. In particular it may be necessary to undertake R&D in order to be able to absorb the implications of other people's R&D, as Cohen and Levinthal (1989) emphasise.

Further work on the costs of acquiring knowledge and the role of external and agglomeration economies would be of great value. We appear to be moving into a phase of development where small venture capital firms based in clusters for biotechnology and information technology seem to be important sources for the creation of knowledge, and we do not fully understand these processes. Studies of diffusion such as that by Stoneman and Kwon in this volume are of great value, and we would again argue that a combination of case study and econometric techniques would be very revealing in this area.

In order that we can further understand the growth of the knowledge stock, rather than its utilisation through innovation, we have to improve our understanding of the process of invention. Much research in the sciences is driven by the process of peer review and approval rather than by the relatively short-term horizon of the market. In this situation it is for the funders of the research, in most cases the state, to determine the underlying direction in which it should go. It is not that they should be 'picking winners', as they are unlikely to be able to do that. Rather they should be deciding which broad areas should be allowed to expand on the basis of evidence of which types of research and education add most to welfare. We need careful work on the effects of R&D, multinational firms and the education system on the process of absorption and creation of knowledge. We also need to know whether R&D will raise the level of output, and in what sectors it is most useful. Econometric work on these topics should be encouraged, but some attention to investment in improving the quality of data would also be important. We have difficulty measuring output in many sectors and we have limited information on innovations and their significance, although work by SPRU and by Hall, both discussed in this volume, has helped us in this respect.

The current government in the UK does not see its role as entirely passive in creating a better environment for production and innovation. There have been clear failures in the UK in these areas as well as successes. The UK has seen a relative decline in its spending on R&D, in part because the government has reduced funds in the area but also because education, especially at the tertiary level, has not been designed to produce researchers in applied sciences. Between 1986 and 1996, business sector R&D, the mainspring of innovation in a modern economy, fell from 1.5 per cent to 1.2 per cent of GDP, whilst in France it rose from 1.2 per cent to 1.5 per cent of GDP. It was considerably higher than this throughout the same period in Germany and the US. The present government thinks it important that this failure of the previous government be addressed as part of a policy on productivity, and the research reported in this book supports that view.

Notes

1 We would like to thank the Economic and Social Research Council under grant R/451/26/441295 for financial support for a conference on Productivity and Competitiveness at which these papers were presented.
2 Griliches (1994) suggests all other sectors are 'unmeasurable'. This position is somewhat extreme in that it suggests that work by national accountants on other sectors is of little value, whereas we argue that improvements in techniques should allow us to begin to solve measurement problems.
3 The market economy excludes education, health and government services.
4 It is important to utilise firm-level data in this area, and this places some restrictions on the timespan available for this research. Firms in the UK have only had to report their R&D expenditure since 1989, although large R&D spenders had been recording it for some time. There is need to combine firm-level and sector-level data in this area, in part because some relevant information, such as survey evidence on skill shortages, is only available at the sector level.
5 However, we must be careful to avoid the association of increases in profits with increased output, but if the activity becomes less profitable then less will be done, and hence there will be fewer spillover effects and less impact on output.

References

Abramovitz, M. (1989), *Thinking About Growth: And Other Essays on Economic Growth and Welfare*, Cambridge, Cambridge University Press.
Aghion, P. and Howitt, P. (1992), 'A model of growth through creative destruction', *Econometrica*, 60, 2, pp. 323–51.
(1998), *Endogenous Growth Theory*, Cambridge, Mass., MIT Press.
Archibugi, D. and Michie, J. (1997), *Technology, Globalisation and Economic Performance*, Cambridge, Cambridge University Press.
Barrell, R. and Pain, N. (1997), 'European integration and foreign direct investment: the UK experience', *National Institute Economic Review*, 160, April.
Barro, R. (1991), 'Economic growth in a cross-section of countries', *Quarterly Journal of Economics*, 106, pp. 407–43.
Barro, R. J. and Sala-i-Martin, X. (1995), *Economic Growth*, New York, McGraw-Hill.
Baumol, W. (1986), 'Productivity growth, convergence and welfare: what the long run data show', *American Economic Review*, 76, pp. 1072–159.
Baumol, W., Blackman, S. and Wolff, E. (1989), *Productivity and American Leader-*

ship, Cambridge, Mass., MIT Press.
Benhabib, J. and Spiegel, M. (1994), 'The role of human capital in economic development: evidence from aggregate cross-country data', *Journal of Monetary Economics*, 34, pp. 143–73.
Braunerhjelm, P. and Ekholm, K. (1999), 'Foreign activities by Swedish multinational corporations', in Barrell, R. and Pain, N. (eds), *Innovation, Investment and the Diffusion of Technology in Europe*, Cambridge, Cambridge University Press.
Broadberry, S. (1994), 'Technological leadership and productivity leadership in manufacturing since the Industrial Revolution: implications for the convergence debate', *Economic Journal*, 104, March.
Caselli, F. (1999), 'Technological revolutions', *American Economic Review*, 89, 1, pp. 78–102.
Chandler, A. (1992), 'Organisational capabilities and the economic history of the industrial enterprise', *Journal of Economic Perspectives*, 6, 3.
Cohen, W. and Levinthal, D. (1989), 'Innovation and learning: the two faces of R&D', *Economic Journal*, 99, pp. 569–86.
Cowing, T.G. and Stevenson, R.E. (1981), *Productivity Measurement in Regulated Industries*, New York, Academic Press.
Crafts, N.F.R. (1998), 'East Asian growth before and after the crisis', IMF Working Paper No. 98/137.
Crafts, N. and Toniolo, G. (1996) (eds), *Economic Growth in Europe Since 1945*, Cambridge, Cambridge University Press.
David, P.A. (1975), *Technical Choice, Innovation and Economic Growth*, Cambridge, Cambridge University Press.
Denison, E.F. (1967), *Why Growth Rates Differ*, Washington, DC, Brookings Institute.
Diewert, W.E. (1976), 'Exact and superlative index numbers', *Journal of Econometrics*, 4, pp. 115–46.
Dixit, A.K. and Stiglitz, J.E. (1977), 'Monopolistic competition and optimum product diversity', *American Economic Review*, 67, pp. 297–308.
Dosi, G. (1988), 'Sources, procedures and microeconomic effects of innovation', *Journal of Economic Literature*, 36, pp. 1126–71.
Dougherty, C. and Jorgenson, D. (1997), 'There is no silver bullet: investment and growth in the G7', *National Institute Economic Review*, 162, October.
Finegold, D. and Soskice, D. (1988), 'The failure of training in Britain', *Oxford Review of Economic Policy*, 4, 3, pp. 21–53.
Freeman, C. (1974), *Economics of Industrial Innovation*, Harmondsworth, Penguin Books.
 (1983) (ed.), *Long Waves in the World Economy*, London, Butterworths.
 (1996), 'The "national system of innovation" in historical perspective', reproduced in Archibugi and Michie (1997).
Freeman, C. and Soete, L. (1987) (eds), *Long Waves and Full Employment*, Oxford, Oxford University Press.
Fujita, M., Krugman, P. and Venables, A. (1999), *The Spatial Economy: Cities, Regions and International Trade*, Cambridge, Mass., MIT Press.

Griliches, Z. (1961), 'Hedonic price indices for automobiles: an econometric analysis of quality change', in *The Price Statistics of the Federal Government*, Washington, NBER.
(1994), 'Productivity, R&D and the data constraint', *American Economic Review*, 84, pp. 1–23.
(1998), *R&D and Productivity: The Econometric Evidence*, Chicago, University of Chicago Press.
Grossman, G.M. and Helpman, E. (1991), *Innovation and Growth in the Global Economy*, Cambridge, Mass., MIT Press.
Howitt, P. and Aghion, P. (1998), 'Capital accumulation and innovation as complementary factors in long-run growth', *Journal of Economic Growth*, 3, pp. 111–30.
Islam, N. (1995), 'Growth empirics: a panel data approach', *Quarterly Journal of Economics*, 110, 4, pp. 1127–70.
Jaffe, A. (1986), 'Technological opportunity and spillovers of R&D', *American Economic Review*, 75, 6, pp. 984–1002.
Jeremy, D.J. (1992) (ed.), *The Transfer Of International Technology: Europe, Japan and the USA in the Twentieth Century*, Aldershot, Edward Elgar.
Jorgenson, D.W., Gollop, F.M. and Fraumeni, B. (1987), *Productivity and US Economic Growth*, Cambridge Mass., Harvard University Press.
Jorgenson, D.W. and Griliches, Z. (1967), 'The explanation of productivity change', *Review of Economic Studies*, 34, 3, pp. 249–83.
Katz, L. and Murphy, K.M. (1992), 'Changes in relative wages, 1963–1987: supply and demand factors', *Quarterly Journal of Economics*, 108, pp. 35–78.
Landau, R. and Rosenberg, N. (1992), 'Successful commercialisation in the chemical process industries', in Landau, R., Mowery, D. and Rosenberg, N. (eds), *Technology and the Wealth of Nations*, Stanford, Stanford University Press.
Landes, D.S. (1969), *The Unbound Prometheus: Technological Change and Industrial Development in Western Europe from 1750 to the Present*, Cambridge, Cambridge University Press.
Levine, R. and Renelt, D. (1992), 'A sensitivity analysis of cross-country growth regressions', *American Economic Review*, 82, 4, pp. 942–63.
Lucas, R.E. (1988), 'On the mechanics of economic development', *Journal of Monetary Economics*, 22, 1, pp. 3–42.
Machin, S. (1998), 'Recent shifts in wage inequality and the wage returns to education in Britain', *National Institute Economic Review*, 166, pp. 87–96.
Maddison, A. (1982), *Phases of Capitalist Development*, Oxford, Oxford University Press.
(1995), *Monitoring the World Economy, 1820–1992*, Paris, Organisation for Economic Co-operation and Development.
Mankiw, N.G., Romer, P., and Weil, D. (1992), 'A contribution to the empirics of economic growth', *Quarterly Journal of Economics*, May, pp. 407–37.
Mason, G. and Finegold, D. (1997), 'Productivity, machinery and skills in the United States and Western Europe', *National Institute Economic Review*, 162, October.
McCloskey, D. (1971), *Essays on a Mature Economy: Britain after 1840*, London, Methuen.

Nelson, R.R. and Winter, S.G. (1982), 'An evolutionary theory of economic change', Cambrige, Mass., Harvard University Press.

Nelson, R. and Wright, G. (1992), 'The rise and fall of American technological leadership: the postwar era in historical perspective', *Journal of Economic Literature*, 30, December.

O'Mahony, M. (1999), *Britain's Productivity Performance, 1950–1996: An International Perspective*, London, National Institute of Economic and Social Research.

Olsen, M. (1996), 'The varieties of Eurosclerosis: the rise and decline of nations since 1982', in Crafts and Toniolo (1996).

Paige, D. and Bombach, G. (1959), *A Comparison of National Output and Productivity of the United Kingdom and the United States*, Paris, Organisation for European Economic Cooperation.

Paque, K.H. (1996), 'Why the 1950s and not the 1920s: Olsonian and non-Olsonian interpretations of 2 decades of German economic history', in Crafts and Toniolo (1996).

Prais, S. (1988), 'Two approaches to the economics of education: a methodological note', *Economics of Education Review*, 7, 2.

(1995), *Productivity, Education and Training: An International Perspective*, Cambridge, Cambridge University Press.

Rodrik, D. (1998), 'Where did all the growth go?: external shocks, social conflict and growth collapses', NBER Working Paper No. 6350, January.

Romer, P.M. (1986), 'Increasing returns and long-run growth', *Journal of Political Economy*, 95, 5, pp. 1002–37.

(1990), 'Endogenous technological change', *Journal of Political Economy*, 98, 5, part II, pp. S71–S102.

(1993), 'Idea gaps and object gaps in economic development', *Journal of Monetary Economics*, 32, pp. 543–73.

Rostas, L. (1948), *Comparative Productivity in British and American Industry*, Cambridge, Cambridge University Press.

Schultz, T.W. (1953), *The Economic Organisation of Agriculture*, New York, McGraw-Hill.

(1961), 'Investment in human capital', *American Economic Review*, 51, pp. 1–17.

(1975), 'The value of the ability to deal with disequilibria', *Journal of Economic Literature*, 31, pp. 199–225.

Schumpeter, J.A. (1934), *The Theory of Economic Development*, Cambridge, Mass., Harvard University Press; Cambridge, Cambridge University Press.

Solow, R.M. (1956), 'A contribution to the theory of economic growth', *Quarterly Journal of Economics*, pp. 65–94.

(1957), 'Technical change and the aggregate production function', *Review of Economics and Statistics*, pp. 312–20.

Spence, M. (1973), 'Job market signaling', *Quarterly Journal of Economics*, 87, pp. 355–74.

Summers, R. and Heston, A. (1991), 'The Penn World Table (Mark 5): an expanded set of international comparisons, 1950–1988', *Quarterly Journal of Economics*, 106, 2, pp. 327–68.

2 Britain's productivity performance in international perspective, 1870–1990: a sectoral analysis

STEPHEN BROADBERRY

1 Introduction

Until recently, data on comparative levels of labour productivity in the advanced industrialised countries for the period since 1870 have been available only at the aggregate level, collected together in an important series of books by Maddison (1964, 1982, 1991, 1995). However, recent research by Broadberry (1993, 1994a, 1994b, 1997a) has provided estimates of comparative productivity levels in manufacturing, and more recently still, in other sectors of the economy (Broadberry, 1997b, 1997c). This paper uses the resultant sectoral breakdown of comparative productivity levels and trends to provide an outline of the ways that the United States and Germany overtook Britain, having begun the period with lower levels of aggregate labour productivity.

The sectoral patterns suggest mechanisms of catching-up and forging ahead that are rather different from those found in the conventional literature. Both Germany and the United States overtook Britain in terms of aggregate labour productivity largely by shifting resources out of agriculture and improving comparative productivity in services rather than by improving comparative productivity in manufacturing. Although capital played some role, the changes in comparative labour productivity also reflected changes in comparative total factor productivity, related to technology and organisation. These results confirm the findings of Bernard and Jones (1996), obtained for the much shorter time period 1970–87, that convergence in aggregate productivity among OECD countries has been driven by services rather than manufacturing. If we are to understand these trends, then, we need to pay less attention to developments within manufacturing, and more attention to wider issues of economic organisation in services.

2 Labour productivity performance by sector, 1870–1990

Tables 2.1 and 2.2 present estimates of comparative labour productivity levels for the US/UK and Germany/UK cases on a sectoral basis over the period 1870–1990. Given the space constraints, it is impossible to do more than sketch the principles underlying the derivation of the estimates, and the reader is referred to Broadberry (1997b, 1997c) for further details. Time series for output and employment in each country are available at the aggregate level and by sector from official and historical national accounts sources.[1] Hence it is possible to construct indices of comparative productivity levels on a sectoral basis. The comparative levels of productivity are then pinned down using a cross-sectional benchmark for each two-country comparison. For the US/UK comparison, for example, this involves comparing the levels of labour productivity in 1937 using physical output per employee or converting value added per employee at a sector-specific price ratio on a purchasing power parity adjusted basis. Since other benchmark estimates are also available for other years, these have been used to provide cross-sectional checks on the time series extrapolations. This means that time series extrapolation over long periods without corroboration can be avoided.

At the whole economy level, we see in table 2.1 that the United States overtook Britain as the labour productivity leader during the 1890s and continued to forge ahead to the 1950s. Since then, there has been a slow process of British catching-up, but there is still a substantial aggregate Anglo-American labour productivity gap of about 30 per cent. Turning to the Germany/UK comparison in table 2.2, the whole economy evidence suggests that German labour productivity in 1870 was only about 60 per cent of the British level, caught up only by the mid-1960s and is now about 20 per cent higher than in Britain. However, the sectoral patterns are a good deal more varied.

2.1 Industrial sectors

One of the most important findings of the sectoral analysis is that there has been a high degree of stability in comparative labour productivity levels in manufacturing, in contrast to the aggregate position, where both the United States and Germany have overtaken Britain. As noted in Broadberry (1993), American labour productivity levels were already about twice the British level in 1870, and the American superiority was still close to this two-to-one level in the late 1980s despite substantial swings in comparative productivity in the intervening years, particularly across the two world

Table 2.1 *Comparative US/UK labour productivity levels by sector, 1869/71–1990 (UK=100)*

	1869/71	1879/81	1889/91	1899/01	1909/11
Agriculture	86.9	98.1	102.1	106.3	103.2
Mineral Extraction	103.1	99.3	109.0	147.3	162.0
Manufacturing	182.5	170.7	193.8	196.5	202.7
Construction	95.5	138.8	164.3	139.7	198.5
Utilities	55.8	74.5	113.5	128.1	149.5
Transport/Comm.	110.0	146.9	167.1	226.8	217.4
Distribution	66.9	107.9	97.0	107.1	120.0
Finance/Services	64.1	58.4	53.2	71.6	77.9
Government	114.3	108.6	102.6	111.2	95.8
GDP	89.8	95.8	94.1	108.0	117.7

	1919/20	1929	1937	1950	1960
Agriculture	128.0	109.7	103.3	126.0	153.1
Mineral Extraction	228.2	248.9	232.1	376.5	618.4
Manufacturing	205.6	250.0	208.3	262.7	243.0
Construction	234.2	133.7	107.8	177.6	235.5
Utilities	295.5	335.9	359.3	573.4	719.9
Transport/Comm.	250.6	231.5	283.4	348.4	318.8
Distribution	109.0	121.9	119.8	135.2	143.2
Finance/Services	103.6	101.5	96.1	111.5	112.3
Government	97.9	99.4	100.0	116.2	110.2
GDP	133.3	139.4	132.6	166.9	167.9

	1968	1973	1979	1990
Agriculture	156.7	131.2	156.1	151.1
Mineral Extraction	700.9	668.0	156.6	119.1
Manufacturing	242.8	215.0	202.6	175.2
Construction	204.5	146.6	129.7	98.5
Utilities	767.9	590.8	523.9	389.8
Transport/Comm.	336.8	303.3	302.7	270.5
Distribution	147.9	149.6	153.8	166.0
Finance/Services	121.3	118.0	118.3	101.0
Government	104.4	101.7	96.5	93.2
GDP	164.2	152.3	145.5	133.0

Source: Broadberry (1997b).
Notes: Extrapolations of output per employee based on 1937 benchmarks.

Table 2.2 *Comparative Germany/UK labour productivity levels by sector, 1871–1990 (UK=100)*

	1871	1881	1891	1901	1911
Agriculture	55.7	54.7	53.7	67.2	67.3
Mineral Extraction	55.9	72.1	80.9	86.4	101.2
Manufacturing	92.6	88.7	94.0	98.8	119.3
Construction	76.1	113.7	90.1	100.3	117.7
Utilities	31.3	49.9	64.2	93.0	103.8
Transport/Comm.	96.8	126.7	147.5	195.1	216.9
Dist'n/Finance	n.a.	38.6	45.9	49.7	52.5
Prof./Pers. Services	89.7	83.4	77.0	76.6	76.3
Government	97.8	95.5	94.6	104.1	98.2
GDP	59.5	57.3	60.5	68.4	75.5

	1925	1929	1935	1950	1960
Agriculture	53.8	56.9	57.2	41.2	47.8
Mineral Extraction	106.8	116.4	123.6	92.4	132.1
Manufacturing	95.2	104.7	102.0	96.0	114.8
Construction	65.7	50.2	70.6	84.2	102.0
Utilities	146.2	158.6	144.0	120.6	151.2
Transport/Comm.	140.0	151.2	132.4	122.0	117.0
Dist'n/Finance	47.1	50.3	54.3	50.7	64.2
Prof./Pers. Services	86.7	99.8	105.6	94.2	85.7
Government	100.1	100.0	100.0	96.9	111.8
GDP	69.0	74.1	75.7	74.4	94.5

	1968	1973	1979	1985	1990
Agriculture	48.6	50.8	65.5	62.1	75.4
Mineral Extraction	135.3	138.1	45.1	25.0	17.5
Manufacturing	120.0	118.6	140.3	121.7	108.3
Construction	105.5	117.7	130.2	111.8	117.9
Utilities	146.7	139.2	164.5	142.7	130.0
Transport/Comm.	130.0	119.5	135.0	132.7	125.7
Dist'n/Finance	75.4	88.0	106.4	109.2	111.2
Prof./Pers. Services	101.3	98.4	103.1	105.3	120.5
Government	111.0	113.3	109.9	111.2	108.6
GDP	107.1	114.0	126.5	120.9	125.4

Source: Broadberry (1997c).
Notes: Extrapolations of output per employee based on 1935 benchmarks.

wars. For the Germany/UK case, we see that in manufacturing, German labour productivity levels were already broadly equal to British levels in the late nineteenth century and remain roughly similar today.

The real contrast in manufacturing, then, has been between Britain and Germany on the one hand, and the United States on the other hand (Broadberry, 1993). The key reason for this transatlantic productivity difference in manufacturing was the coexistence of two technological systems, geared around 'mass production' and 'flexible production' (Piore and Sabel, 1984, Tolliday and Zeitlin, 1991, Broadberry, 1994a). In mass production, special purpose machinery was substituted for skilled shopfloor labour to produce standardised products, while flexible production relied on skilled shopfloor labour to produce customised output. Although it would be an over-simplification to identify American manufacturing with mass production and European manufacturing with flexible production, since in practice both systems coexisted on both sides of the Atlantic, mass production has been more prevalent in America and flexible production more prevalent in Britain and Germany, for reasons associated with both demand and supply conditions.[2] On the demand side, standardisation was facilitated in the United States by the existence of a large homogeneous home market, compared with the fragmentation of national markets and greater reliance on exports in Europe, while on the supply side, mass production machinery economised on skilled shopfloor labour (relatively abundant in Europe) but was wasteful of natural resources (relatively scarce in Europe). For a thorough discussion of these issues, including a detailed analysis of individual industries, the reader is referred to Broadberry (1997a).

Construction exhibits inverse long swings or Kuznets cycles in both the US/UK and Germany/UK cases (Thomas, 1973). In the US/UK case, British labour productivity has been broadly equal to American labour productivity when the British industry was in a boom and the US industry in a slump, while the US industry had a two-to-one labour productivity lead when the British industry was in a slump and the US industry in a boom. For the Germany/UK case, German productivity has been higher when the German industry was booming, and British labour productivity has been higher when the British industry was booming.

In mineral extraction we see the importance of natural resources and composition effects. In the United States, we see the growing importance of the high value added oil industry through much of the twentieth century, although the establishment of the North Sea oil industry in Britain has dramatically improved Britain's productivity performance in this sector since 1973 (Kendrick, 1961 p. 375, Bean, 1987). The rundown of the low productivity deep-mined coal industry in Britain has reinforced this improvement (Glyn and Machin, 1996).

Developments in the utilities sector also reflect output composition effects and resource discoveries. The shift away from manufactured gas to natural gas and electricity (where the US has an advantage from the more readily exploitable hydroelectric resources) greatly increased the scale of the American labour productivity lead during much of the twentieth century (Gould, 1946 pp. 55, 94–96, Hannah, 1979 p. 130). The discovery of North Sea gas has allowed Britain to close the gap since the early 1970s.

2.2 Agriculture and structural change

The position in agriculture is that British performance has been better than in industrial sectors when assessed on a levels basis. However, whereas there has been no deterioration over the long run in Britain's comparative labour productivity performance in manufacturing, there has been a trend deterioration in agriculture: the United States has pulled ahead in the twentieth century. The trend is less clear in the Germany/UK case because of German setbacks across the two world wars, but the German position by the 1980s had improved relative to the German position of the 1870s. The high level of labour productivity in British agriculture during the late nineteenth century can be largely explained by the high level of capital per worker and the concentration on high value added livestock products, which were themselves results of the importation of cheap grain from the New World (Rostas, 1948 p. 80). Labour intensive arable production was ruled out in Britain by free trade in corn.

The contribution of agriculture to American and German overtaking of Britain was mainly through the later release of labour in the former two countries. Data on the structure of employment in the three countries since 1870 are given in table 2.3. Whereas in about 1870 agriculture accounted for less than a quarter of the labour force in Britain, in the United States and Germany it accounted for about a half. With higher value added per employee in industry and services than in agriculture, there was much greater scope in Germany and the United States for favourable effects from the sectoral reallocation of labour. As recent research on the Industrial Revolution has shown, Britain's achievement in the nineteenth century was not really the creation of a high productivity industrial sector so much as a large industrial sector that was actually quite labour intensive (Crafts, 1985 p. 156). And this creation of a large industrial sector depended on the early release of labour from agriculture.

Table 2.3 *Sectoral shares of employment in the United States, the United Kingdom and Germany, 1870–1990 (%)*

A. United States	1870	1910	1930	1950	1990
Agriculture	50.0	32.0	20.9	11.0	2.5
Mineral Extraction	1.5	2.8	2.2	1.5	0.6
Manufacturing	17.3	22.2	21.3	25.0	15.3
Construction	5.8	6.3	5.9	5.5	5.2
Utilities	0.2	0.5	0.8	0.9	0.7
Transport/Comm.	4.6	8.1	8.6	6.0	4.0
Distribution	6.1	9.1	11.7	18.7	22.0
Finance/Services	12.2	17.1	21.4	21.3	40.2
Government	2.3	1.9	7.2	10.1	9.5
Total	100.0	100.0	100.0	100.0	100.0

B. United Kingdom	1871	1911	1930	1950	1990
Agriculture	22.2	11.8	7.6	5.1	2.0
Mineral Extraction	4.0	6.3	5.4	3.7	0.6
Manufacturing	33.5	32.1	31.7	34.9	20.1
Construction	4.7	5.1	5.4	6.3	6.7
Utilities	0.2	0.6	1.2	1.6	1.1
Transport/Comm.	5.4	7.7	8.3	7.9	5.5
Distribution	7.5	12.1	14.3	12.2	19.5
Finance/Services	19.5	20.2	20.9	19.5	37.5
Government	3.0	4.1	5.2	8.8	7.0
Total	100.0	100.0	100.0	100.0	100.0

C. Germany	1875	1913	1935	1950	1990
Agriculture	49.5	34.5	29.9	24.3	3.4
Mineral Extraction	1.5	2.8	1.7	2.8	0.6
Manufacturing	24.7	29.5	30.0	31.4	31.4
Construction	2.8	5.3	5.9	7.2	6.7
Utilities	0.1	0.3	0.6	0.7	1.0
Transport/Comm.	1.9	3.8	4.8	5.6	5.6
Dist'n/Finance	6.0	11.2	13.5	13.2	16.3
Prof./Pers. Services	10.0	8.3	8.8	7.9	19.9
Government	3.5	4.3	4.8	6.9	15.1
Total	100.0	100.0	100.0	100.0	100.0

Source: Broadberry (1998a).

2.3 Services

Like agriculture, British service sectors have generally shown a better performance than industrial sectors if we consider levels of labour productivity. But again, as in agriculture, there has been a trend deterioration of Britain's comparative productivity performance in service sectors. Although the productivity gap has been larger in manufacturing, it has not grown bigger over time; in services, by contrast, Britain has moved from a position of a sizeable productivity lead to a small productivity shortfall. We shall focus on this change in comparative productivity performance in services in section 4. Before that, however, we need to consider the contribution of capital and derive estimates of comparative total factor productivity levels.

2.4 Summary of labour productivity patterns at the sectoral level

Most studies of economic growth jump from evidence of comparative productivity performance at the aggregate level to theories centred on technology in manufacturing, on the presumption that there is a single best-practice technology that would be applied throughout the world if it were not for barriers to diffusion.[3] However, the detailed historical evidence from the three countries considered here raises doubts about this interpretation on at least two counts. First, large labour productivity gaps have persisted in manufacturing between the old world and new world countries, which can be related to persistent differences in technology, despite access to a common pool of knowledge. And second, the changing aggregate levels of comparative labour productivity owe more to developments in services and the shift of resources out of agriculture than to developments in manufacturing.

3 Capital stocks and total factor productivity

Much of the literature on international comparisons suggests that labour productivity growth rates differ at least in part because of differences in capital stock growth (Denison, 1967, Maddison, 1987). Hence it is of some interest to consider the role of capital in explaining differences in labour productivity levels. Working in terms of levels provides a check on the consistency and plausibility of trends in the growth rate of capital stocks in the three countries. This is a serious issue, because concerns have been expressed about the international comparability of capital stock estimates derived from data on investment using official asset life assumptions that vary substantially between countries (Maddison, 1995, O'Mahony, 1996).

Detailed sources for the basic capital stock time series are given in Broadberry (1998a). Comparative levels of TFP for the US/UK case are given in table 2.4. Comparing these comparative TFP levels with the comparative labour productivity levels in table 2.1, we see that capital has a role to play in explaining labour productivity differences, but not enough to eliminate TFP differences altogether. Note, however, that at the aggregate level, whereas the United States overtook Britain in terms of labour productivity before the turn of the century, it was only between the wars that the United States gained a TFP advantage. This is consistent with the emphasis of Abramovitz and David (1973, 1994) on the role of capital in American economic growth during the nineteenth century. It is also consistent with McCloskey's (1970) claim that Victorian Britain did not fail, in the sense that the United States was still catching-up in terms of aggregate TFP levels. Turning to the sectoral data, in agriculture capital per employee has grown faster in the United States than in Britain, so that US/UK comparative TFP has grown by less than comparative labour productivity. In mineral extraction and also in the utilities, the rise and fall of the US TFP lead is damped relative to the labour productivity lead because of a large swing in comparative capital intensity. In transport and communications, higher capital per employee in the United States reduces the American TFP lead compared with the labour productivity lead, although the American capital intensity advantage diminishes over time. In manufacturing, however, capital per employee has been broadly similar in the two countries, so that labour productivity and TFP gaps remain broadly similar, while in distribution the US TFP lead is wider than the labour productivity lead, since capital intensity is higher in Britain.

Owing to the limitations of the German historical capital stock data, comparative levels of TFP for the Germany/UK case in table 2.5 are less complete than for the US/UK case. Comparing these comparative TFP levels with the comparative labour productivity levels in table 2.2, we can see that in the Germany/UK case, as in the US/UK case, capital has some role to play in explaining labour productivity differences, but not enough to eliminate TFP differences altogether. At the aggregate level, with capital per employee growing faster in Germany than in Britain, German overtaking has been less marked in terms of TFP than in terms of labour productivity. On a sectoral basis, long-term estimates are only available for agriculture and manufacturing. In agriculture, the rapid growth of capital per employee in Germany means that there has been no catching-up in terms of TFP. In manufacturing, although capital per employee has

Table 2.4 *Comparative US/UK total factor productivity levels by sector (UK=100)*

	1869/71	1879/81	1889/91	1899/01	1909/11
Agriculture	99.7	115.5	123.1	126.8	118.8
Mineral Extraction	89.8	78.7	82.7	104.7	113.5
Manufacturing	193.7	180.3	180.3	177.5	180.5
Construction					
Utilities	73.3	79.3	99.6	101.1	126.5
Transport/Comm.				180.2	180.5
Distribution					
Finance/Services					
Government					
GDP	91.9	92.5	84.7	90.2	95.2

	1919/20	1929	1937	1950	1960
Agriculture	134.9	121.1	125.1	136.7	145.5
Mineral Extraction	147.5	158.9	167.0	246.2	378.1
Manufacturing	183.2	227.0	188.8	249.8	227.1
Construction					
Utilities	230.1	274.3	292.1	406.7	509.6
Transport/Comm.	208.5	196.9	228.2	290.3	274.4
Distribution		132.2	139.9	154.5	167.3
Finance/Services					
Government					
GDP	105.9	111.4	106.0	133.0	134.2

	1968	1973	1979	1990
Agriculture	143.4	124.8	144.1	136.9
Mineral Extraction	451.0	461.1	150.1	134.1
Manufacturing	237.3	211.5	202.3	175.2
Construction				
Utilities	579.4	489.1	424.7	320.9
Transport/Comm.	301.3	280.5	282.7	247.9
Distribution	183.9	190.5	204.2	225.9
Finance/Services				
Government				
GDP	138.4	132.6	131.4	121.4

Source: Broadberry (1998a).

Table 2.5 *Comparative Germany/UK total factor productivity levels by sector (UK=100)*

	1871	1881	1891	1901	1911
Agriculture	58.4	60.9	59.8	73.7	71.7
Mineral Extraction					
Manufacturing	111.9	109.9	106.7	104.9	124.7
Construction					
Utilities					
Transport/Comm.					
Dist'n/Finance					
Prof./Pers. Services					
Government					
GDP	60.9	60.2	62.4	68.7	75.3

	1925	1929	1935	1950	1960
Agriculture	58.1	60.2	60.5	46.9	47.9
Mineral Extraction				84.6	115.1
Manufacturing	114.3	121.3	115.9	108.4	126.1
Construction					
Utilities				120.6	151.2
Transport/Comm.				139.9	136.3
Dist'n/Finance				70.8	82.7
Prof./Pers. Services					
Government					
GDP	74.1	78.3	78.0	76.4	95.0

	1968	1973	1979	1985	1990
Agriculture	44.6	46.6	56.9	52.0	60.2
Mineral Extraction	116.8	124.3	52.2	32.1	24.3
Manufacturing	123.8	121.2	143.2	131.8	119.4
Construction					
Utilities	146.7	139.2	164.5	142.7	130.0
Transport/Comm.	144.8	132.7	145.3	139.7	130.0
Dist'n/Finance	95.6	112.1	138.1	145.1	154.7
Prof./Pers. Services					
Government					
GDP	101.4	107.7	117.2	111.6	114.7

Source: Broadberry (1998a).

grown more rapidly in Germany, the level of capital intensity has nevertheless remained below the British level, so that Germany has maintained a TFP lead throughout the period. For the post-1950 period, faster growth of capital intensity in Germany dampens German TFP performance compared with labour productivity performance in the utilities and in transport and communications. In mineral extraction, capital intensity rose faster in Germany to 1968, then faster in Britain, so that the swing in TFP is more muted than in labour productivity. However, in distribution and finance, German overtaking in terms of labour productivity has occurred without a turnaround in relative capital intensity, so that German TFP performance appears even more impressive.

Although Maddison (1995) argues for standardisation of asset lives in international comparisons, there are serious practical objections. The first problem is that a substantial proportion of the historical capital stock data is based on direct estimates of the capital stock from fire insurance records and censuses, which are used to obtain estimates of investment, thus inverting the usual perpetual inventory method. To use these investment data to obtain new estimates of the capital stock would inevitably involve some circularity of argument. The second problem is that Maddison's (1995) application of the standardisation procedure at the aggregate level produces an implausibly large US/UK capital intensity lead. For example, Maddison's (1995 pp. 253–4) standardised estimates produce a US/UK capital per employee ratio of more than five to one in 1938, when the United States managed to produce less than 70 per cent more output per employee than Britain. For the long period studied here, then, we are forced to rely on official capital stock data.

The above figures suggest that one way in which the United States and Germany overtook Britain in terms of labour productivity was by investing more. As capital per employee grew faster in the United States and Germany, so Britain fell behind in terms of output per employee. However, this is not the whole story, because there are also differences in TFP. Britain has been overtaken in terms of TFP as well as labour productivity, suggesting that technology and organisation have also been important.

4 Explaining service sector trends

The least understood aspect of Britain's relative economic decline must surely be the loss of labour productivity leadership in services, taken in this study to include transport and communications, distribution, finance, personal and professional services and government. To the extent that services are discussed at all in the debate on comparative economic performance, it is usual to make the point that British performance was not as bad as in manufacturing (Lee, 1986, Gemmell and Wardley, 1990, Rubinstein, 1993). However, this confuses levels with rates of change. Although the productivity gap has been larger in manufacturing, it has not grown bigger over time; in services, by contrast, Britain has moved from a position of a sizeable productivity lead to a small productivity deficit. It should be noted that this is not a statistical illusion; although there are some difficulties of measurement in non-market services, there is plenty of reliable evidence to support these trends in market services such as transport and communications, distribution, financial services and some professional and personal services.[4] In this section we look at the factors underlying Britain's early labour productivity leadership in services, and the reasons for the loss of that leadership.

4.1 Urbanisation and economies of agglomeration

In distribution, finance, professional and personal services, Britain's productivity position before 1914 benefited from the relatively high degree of urbanisation. In distribution, for example, Jefferys (1954 p. 34) sees the rapid growth of the department store, the multiple shop and the co-operative forms of retailing in late-nineteenth-century Britain as dependent on the existence of a large, steady and consistent demand from a relatively homogeneous urban working class. The contrast here is most obvious with respect to Germany, for in the United States, as we have seen, most writers have stressed the homogeneity of demand, even among the rural population. Whereas the mail-order store provided cheap, homogenous goods to the rural consumer in the United States, this occurred on a much more limited scale in Germany (Chandler, 1990 pp. 59, 420).

Similarly, the higher density of financial and other commercial institutions in late-nineteenth-century Britain can be understood in relation to the higher urban density, with concentrated urban demands allowing a high degree of specialisation (Smolensky, 1972). Clearly, however, such advantages were not sustainable in the long run as the United States and Germany industrialised and urbanised. The process proceeded more slowly in Germany, where as late as 1950, almost a quarter of the labour force was still engaged in agriculture.

4.2 Networks and hierarchies

Although international comparisons are much less developed in service sectors than in industrial sectors, the literature does nevertheless point to the establishment of networks as one source of British success before the First World War. Boyce (1995 pp. 3–4) sees networks of individuals bound together by interpersonal knowledge as reducing the threat of opportunistic behaviour in a world of asymmetric information. Repeat contracting based on reciprocity generated trust and co-operation; reputation was accumulated only gradually, but could be destroyed immediately by opportunistic behaviour, which was thus deterred. Reputation could gradually become shared as successful co-operation within a network reflected favourably on all participants. Hence network membership tended to be exclusive because opportunistic newcomers could destroy the collective reputation of a network. To some extent, then, the social exclusivity of British business networks can be seen as a source of success in sectors that are characterised by non-contractibility and customised transactions, and not just as a source of weakness in sectors characterised by standardised transactions and hierarchy, as in the more traditional literature on British economic decline. The contrast between these two views of British culture can be seen in the pro-service interpretation of Rubinstein (1993) and the anti-industry interpretation of Wiener (1981).

Over time, however, the advantages of the network form of organisation became outweighed by the disadvantages, and an alternative form of organisation based on hierarchy provided a serious alternative. This form of organisation was particularly prevalent in the United States, and has been identified as a major source of US competitive advantage by Chandler (1980). However, whereas Chandler (1990) concentrates on the emergence of the large scale hierarchical corporation in manufacturing, his earlier work emphasised the role of a number of service sectors, including the railways and distribution (Chandler, 1977). The key factors underlying the growth of hierarchical forms of organisation in the service sector were, first, developments in communications technology, reducing problems of asymmetric information and allowing much closer contact between principal and agent in merchant/financial operations (Nishimura, 1971, Chapman, 1992 pp. 193–230); and second, the growing volume of economic activity, permitting greater specialisation in services and hence allowing task simplification and easier monitoring of employee performance. Nevertheless, the extent to which networks remained competitive varied between sectors, and British performance tended to be better in sectors where conditions continued to favour networks over hierarchy. Indeed, Britain has retained a strong position to this day in international financial services on the basis of the City of London's networks (Smith, 1992).

4.3 Human capital

Another aspect often stressed in more optimistic service based accounts of British economic performance is the early emergence in Britain of professional associations, which increasingly set standards for training and conducted professional examinations, as well as limiting opportunistic behaviour in areas characterised by asymmetric information, such as law and medicine (Reader, 1966). If we take seriously the modern distinction between higher level and intermediate level qualifications, and if we are prepared to treat membership of professional associations as a higher level qualification, then this gave Britain a human capital advantage (Prais, 1995, Broadberry, 1998b). Many of these qualified professionals worked in the commercial sectors, where Britain had a labour productivity lead over both Germany and the United States before the First World War. Furthermore, a high proportion of these professional qualifications has been in accountancy and other commercial areas, so that Cassis (1997 p. 139) argues that British business élites had a more commercially oriented education than in other European countries.

However, again this early British human capital advantage has not been sustained. Indeed, it may even be argued that Britain has fallen behind in this area with the emergence of a US advantage at the higher level based on university training, and a German advantage at the intermediate level based on vocational training, with the growing provision of apprenticeships in the service sector as well as in industry (Broadberry, 1998b). However, we should recognise the dangers of excessive credentialism, particularly in service sectors where the main skill requirements are often interpersonal rather than cognitive or motor skills (Howell and Wolff, 1992 p. 128, Gallie and White, 1993).

4.4 International economic relations

During the period covering the two world wars there was a major change in the climate of international economic relations, with the disintegration of the liberal world economic system. This provided particular difficulties for Britain, where economic activity had been highly globalised before the First World War. The response was a move towards greater integration within the British Empire, building on some of the networks established before 1914 (Drummond, 1974). Whilst this may have been beneficial in the short run, limiting the impact of the depression of the 1930s, it created further problems of adjustment with the reintegration of the liberal world economic system after the Second World War

(Broadberry, 1997a). The geographical specificity of network relations meant that many British enterprises were effectively 'locked in' to markets which were growing slowly and in which Britain had no natural competitive advantage (Jones, 1993 pp. 285–94).

4.5 Lock-in effects

Lock-in can also occur because of prior accumulation of physical capital, and the British railway sector provides one of the most well known examples of this. Given the interrelatedness of the capital stock, the early start in Britain meant that the railway system was geared around technology that was becoming increasingly obsolete, and piecemeal investments were often unable to embody best practice. As Clapham (1938 p. 350) notes, the size of locomotives, coaches and wagons was limited by 'bridges in congested towns, the short radius of curves and the whole lay-out of stations, docks and staithes'. The example of what Veblen (1915 p. 130) famously described as the 'silly little bobtailed' wagons for moving coal on British railways was used by Frankel (1955) in his formalisation of the problem of technical interrelatedness and investment in a mature economy. In fact, van Vleck (1997 p. 140) has recently disputed this particular case, arguing that the wagons were small because this provided flexibility and substituted for the more expensive means of delivering coal by road. Nevertheless, even if this point is conceded, the general argument remains valid for other types of rolling stock and locomotives.

5 Concluding comments

A sectoral analysis of comparative labour productivity levels over the period since 1870 for Britain, the United States and Germany reveals a rather different pattern from that which appears (at least implicitly) in the textbooks. The United States and Germany caught up with and overtook Britain largely as a result of developments within services, together with a structural transformation away from agriculture. Comparative productivity levels in manufacturing have remained relatively stable. These patterns remain if we examine total factor productivity; fixed capital can explain some of the differences in labour productivity, but there is still a role for technology and organisation.

Notes

1 The key historical national accounts sources are Feinstein (1972) for Britain, Kendrick (1961) for the United States and Hoffmann (1965) for Germany.
2 See Scranton (1991) for a discussion of flexible production in an American context, and Herrigel (1996) for an analysis of German industrial performance stressing the coexistence of mass production and flexible production.
3 Recent work on endogenous growth reinforces this way of thinking, with its emphasis on research and development, an activity almost exclusively related to manufacturing. See, for example, the recent textbooks by Barro and Sala-i-Martin (1995) and Jones (1998).
4 Indicators of comparative labour productivity levels at a more disaggregated level are available in Broadberry (1997b; 1997c).

References

Abramovitz, M. and David, P.A. (1973), 'Reinterpreting economic growth: parables and realities', *American Economic Review*, 63, pp. 428–39.
 (1994), 'Convergence and deferred catch-up', in Landau, R., Taylor, T. and Wright, G. (eds), *The Mosaic of Economic Growth*, Stanford, Stanford University Press, pp. 21–62.
Barro, R.J. and Sala-i-Martin, X. (1995), *Economic Growth*, New York, McGraw-Hill.
Bean, C. (1987), 'The impact of North Sea oil', in Dornbusch, R. and Layard, R. (eds), *The Performance of the British Economy*, Oxford, Oxford University Press, pp. 64–96.
Bernard, A. and Jones, C.I. (1996), 'Comparing apples to oranges: productivity convergence and measurement across industries and countries', *American Economic Review*, 86, pp. 1216–38.
Boyce, G. (1995), *Information, Mediation and Institutional Development: The Rise of Large-Scale Enterprise in British Shipping, 1870–1919*, Manchester, Manchester University Press.
Broadberry, S.N. (1993), 'Manufacturing and the convergence hypothesis: what the long run data show', *Journal of Economic History*, 53, pp. 772–95.
 (1994a), 'Technological leadership and productivity leadership in manufacturing since the Industrial Revolution: implications for the convergence debate', *Economic Journal*, 104, pp. 291–302.
 (1994b), 'Comparative productivity in British and American manufacturing during the nineteenth century', *Explorations in Economic History*, 31, pp. 521–48.
 (1997a), *The Productivity Race: British Manufacturing in International Perspective, 1850–1990*, Cambridge, Cambridge University Press.

(1997b), 'Forging ahead, falling behind and catching-up: a sectoral analysis of Anglo-American productivity differences, 1870–1990', *Research in Economic History*, 17, pp. 1–37.

(1997c), 'Anglo-German productivity differences 1870–1990: a sectoral analysis', *European Review of Economic History*, 1, pp. 247–67.

(1998a), 'How did the United States and Germany overtake Britain? A sectoral analysis of comparative productivity levels, 1870–1990', *Journal of Economic History*, 58, pp. 375–407.

(1998b), 'Human capital and productivity performance in twentieth century Britain: an international perspective', unpublished, University of Warwick.

Cassis, Y. (1997), *Big Business: The European Experience in the Twentieth Century*, Oxford, Oxford University Press.

Chandler, A.D., Jr (1977), *The Visible Hand: The Managerial Revolution in American Business*, Cambridge, Mass., Harvard University Press.

(1980), 'The growth of the transnational industrial firm in the United States and the United Kingdom: a comparative analysis', *Economic History Review*, 33, pp. 396–410.

(1990), *Scale and Scope: The Dynamics of Industrial Capitalism*, Cambridge, Mass., Harvard University Press.

Chapman, S. (1992), *Merchant Enterprise in Britain: From the Industrial Revolution to World War I*, Cambridge, Cambridge University Press.

Clapham, J.H. (1938), *An Economic History of Modern Britain: Machines and National Rivalries (1887–1914), with an Epilogue (1914–1929)*, Cambridge, Cambridge University Press.

Crafts, N.F.R. (1985), *British Economic Growth During the Industrial Revolution*, Oxford, Oxford University Press.

Denison, E.F. (1967), *Why Growth Rates Differ: Postwar Experience in Nine Western Countries*, Washington DC, The Brookings Institution.

Drummond, I.M. (1974), *Imperial Economic Policy, 1917–1939: Studies in Expansion and Protection*, London, Allen & Unwin.

Feinstein, C.H. (1972), *National Income, Expenditure and Output of the United Kingdom, 1855–1965*, Cambridge, Cambridge University Press.

Frankel, M. (1955), 'Obsolescence and technological change in a maturing economy', *American Economic Review*, 45, pp. 296–319.

Gallie, D. and White, M. (1993), *Employee Commitment and the Skills Revolution*, London, Policy Studies Institute.

Gemmell, N. and Wardley, P. (1990), 'The contribution of services to British economic growth, 1856–1913', *Explorations in Economic History*, 27, pp. 299–321.

Glyn, A. and Machin, S. (1996), 'Colliery closures and the decline of the UK coal industry', Discussion Paper Series on the Labour Market Consequences of Technical and Structural Change, No. 7, Centre for Economic Performance, London School of Economics.

Gould, J.M. (1946), *Output and Productivity in the Electric and Gas Utilities, 1899–1942*, New York, National Bureau of Economic Research.

Hannah, L. (1979), *Electricity Before Nationalisation: A Study of the Development*

of the Electricity Supply Industry in Britain to 1948, London, Macmillan.
Herrigel, G. (1996), *Power*, Cambridge, Cambridge University Press.
Hoffmann, W.G. (1965), *Das Wachstum der deutschen Wirtschaft seit der Mitte des 19. Jahrhunderts*, Berlin, Springer-Verlag.
Howell, D.R. and Wolff, E.N. (1992), 'Technical change and the demand for skills by US industries', *Cambridge Journal of Economics*, 16, pp. 127–46.
Jefferys, J.B. (1954), *Retail Trading in Britain, 1850–1950*, Cambridge, Cambridge University Press.
Jones, C. I. (1998), *Introduction to Economic Growth*, London, Norton.
Jones, G. (1993), *British Multinational Banking, 1830–1990*, Oxford, Oxford University Press.
Kendrick, J.W. (1961), *Productivity Trends in the United States*, Princeton, National Bureau of Economic Research.
Lee, C.H. (1986), *The British Economy Since 1700: A Macroeconomic Perspective*, Cambridge, Cambridge University Press.
Maddison, A. (1964), *Economic Growth in the West*, London, Allen & Unwin.
(1982), *Phases of Capitalist Development*, Oxford, Oxford University Press.
(1987), 'Growth and slowdown in advanced capitalist economies: techniques of quantitative assessment', *Journal of Economic Literature*, 25, pp. 649–98.
(1991), *Dynamic Forces in Capitalist Development*, Oxford, Oxford University Press.
(1995), *Monitoring the World Economy, 1820–1992*, Paris, Organisation for Economic Co-operation and Development.
McCloskey, D.N. (1970), 'Did Victorian Britain fail?', *Economic History Review*, 23, pp. 446–59.
Nishimura, S. (1971), *The Decline of the Inland Bills of Exchange in the London Money Market, 1855–1913*, Cambridge, Cambridge University Press.
O'Mahony, M. (1996), 'Measures of fixed capital stocks in the post-war period: a five-country study', in van Ark, B. and Crafts, N.F.R. (eds), *Quantitative Aspects of Post-War European Economic Growth*, Cambridge, Cambridge University Press, pp. 165–214.
Piore, M.J. and Sabel, C.F. (1984), *The Second Industrial Divide: Possibilities for Prosperity*, New York, Basic.
Prais, S.J. (1995), *Productivity, Education and Training: An International Perspective*, Cambridge, Cambridge University Press.
Reader, W. J. (1966), *Professional Men: The Rise of the Professional Classes in Nineteenth-Century England*, London, Weidenfeld and Nicolson.
Rostas, L. (1948), *Comparative Productivity in British and American Industry*, Cambridge, National Institute of Economic and Social Research.
Rubinstein, W.D. (1993), *Capitalism, Culture and Economic Decline in Britain, 1750–1990*, London, Routledge.
Scranton, P. (1991), 'Diversity in diversity: flexible production and American industrialization', *Business History Review*, 65, pp. 27–90.
Smith, A.D. (1992), *International Financial Markets: The Performance of Britain and its Rivals*, Cambridge, Cambridge University Press.

Smolensky, E. (1972), 'Industrial location and urban growth', in Davis, L.E., Easterlin, R.A. and Parker, W.N. (eds), *American Economic Growth: An Economist's History of the United States*, New York, Harper & Row, pp. 582–607.

Thomas, B. (1973), *Migration and Economic Growth: A Study of Great Britain and the Atlantic Economy*, (2nd edn), Cambridge, Cambridge University Press.

Tolliday, S. and Zeitlin, J. (1991) (eds), *The Power to Manage? Employers and Industrial Relations in Comparative Historical Perspective*, London, Routledge.

Veblen, T. (1915), *Imperial Germany and the Industrial Revolution*, London, Macmillan.

van Vleck, V.N.L. (1997), 'Delivering coal by road and rail in Britain: the efficiency of the "silly little bobtailed" wagons', *Journal of Economic History*, 57, pp. 139–60.

Wiener, M.J. (1981), *English Culture and the Decline of the Industrial Spirit, 1850–1990*, Cambridge, Cambridge University Press.

3 Labour productivity and product quality: their growth and inter-industry transmission in the UK 1979–1990

CHRISTINE GREENHALGH AND MARY GREGORY[1]

The motivation and framework for this study

The record on labour productivity growth in the UK in the 1980s is now acknowledged to have been respectable rather than spectacular; better than the 1970s but inferior to the immediate postwar decades, and well short of the 'miracle' claimed for it towards the end of the Thatcher years (Muellbauer, 1991; Bean and Crafts, 1996). Our own earlier work on output and employment change between 1979 and 1990 indicates that economies in labour use almost exactly counterbalanced the employment-creating effects of the growth of output (Gregory and Greenhalgh, 1997). This process of productivity growth/labour-saving has encompassed several different developments. Job-shedding has became a major feature of the UK economy as corporate Britain has 'down-sized' and restructured. In part this reflects process innovation and the accompanying reorganisation at the workplace. A further part, however, reflects the move towards contracting out of activities previously conducted in-house. Out-sourcing in the search for efficiency gains brings increased specialisation and thus creates, destroys and reallocates jobs. The emphasis on down-sizing has been accompanied by a growing awareness that cost-saving alone is not sufficient, and that competitive success increasingly involves the development of new products and services, incorporating innovations and continuous quality improvement. These various dimensions of productivity growth are the focus of our analysis.

Labour productivity growth is most commonly analysed in the context of multi- or total factor productivity growth. Building on the seminal contribution of Solow (1957) the standard vehicle for this type of analysis is the neoclassical production function, often Cobb–Douglas in empirical implementation. Recent examples include Bernard and Jones (1996), Wolff (1996), Hall and Mairesse (1995) and Oulton and O'Mahony (1994). Hart (1996), developing insights from Jorgenson and Griliches (1967) and Griliches (1990), has recently re-emphasised that many of the results on total factor productivity can be derived directly from an accounting framework, without invoking underpinnings from neoclassical factor pricing. Our approach is similarly largely empirical and broadly in the growth accounting tradition, but differs in several key aspects. Our dual focus is on labour productivity and product quality. We view changes in labour productivity as rooted in changes in the organisation of production and in the development of new products rather than in primary factor efficiency (Carter, 1970). This is not to deny the importance of skill acquisition, which is currently being widely addressed in studies of economic growth, but to sharpen the focus on structural change. This twin emphasis is based on the view that, in an open trading economy, labour endowment and the stock of technology underpin comparative advantage, as firms aim to combine capital with raw materials and intermediate goods to minimise costs and raise product quality; this approach echoes both new growth theory and modern trade analysis (Grossman and Helpman, 1996) .

We look at a single economy, the UK, at a sectoral level, with the sectors encompassing the whole economy. This disaggregated perspective recognises not only that labour productivity growth proceeds at varying rates in different sectors but that its transmission across sectors is an intrinsic part of the growth process. This approach echoes an earlier debate in the growth accounting literature on the appropriate methods for analysis of productivity growth in an economy with intermediate products, in which the whole is more than the simple sum of the parts due to intersectoral impacts of technical change (Hulten, 1978). As our results will show, services play a key role both as sources of labour productivity growth and as major elements in its transmission. An exclusive focus on manufacturing, for example, would fail to capture these aspects.

We look at labour productivity growth in parallel with the generation and diffusion of R&D. As with labour cost-saving, the outcomes of R&D can be viewed as embodied in sectoral outputs and transmitted through the economy by inter-industry sales and purchases. The same analytical framework can thus be applied to these two dimensions of productivity gains and their transmission through the economy. Finally, our analysis is based on gross output in place of the more familiar value-added production function. This is the appropriate approach at the sectoral level (Jorgenson, Gollop and Fraumeni, 1987) and reflects our view that the emphasis on labour cost-saving over the 1980s has been part of a much wider cost-saving approach, including contracting-out and just-in-time delivery of materials and other inputs. Similarly, the role of intermediate goods in embodying R&D and product innovations which are then 'bought in' by other sectors is widely recognised. To capture these aspects requires the full range of purchased inputs in addition to primary factors.

Our first objective is to analyse the sources and transmission of labour-saving innovation and technical change in the UK economy, identifying sectors where labour productivity growth has been particularly strong and/or which have had a significant effect, through inter-industry purchases and sales, on other areas of the economy. In parallel with this analysis of the transmission of labour cost-saving, we examine the direct and indirect R&D intensity of production and the inter-industry transmission of the outcomes of R&D expenditure. Taking business R&D as a proxy measure of changes in product quality gives the transmission of product quality improvement between sectors. We can then examine how far the sectors characterised by high growth rates and/or large spillovers of productivity are also sources of high rates of innovation arising from business R&D. By using input–output relationships from different dates we gain indicators of the role of contracting-out. In this way our analysis encompasses three broad sources of productivity growth: process innovation at the workplace, which is frequently job-destroying; product innovation to improve product quality, often seen as employment-creating; and restructuring and the contracting-out of work between sectors, bringing efficiency gains through specialisation.

Our empirical framework is based on the input–output tables for the UK for 1979, 1985 and 1990. These provide coverage of the economy as a whole combined with extensive sectoral disaggregation. The input–output framework has been extensively used to analyse the economy-wide linkages in industry outputs. We extend this methodology to examine the transmission of labour cost-saving and R&D across sectors. The articulation of inter-industry transactions in intermediate goods allows linkages to be traced from final use backwards along supply chains and forwards from suppliers to purchasers. These are channels through which producers may seek cost-savings and quality improvements from their suppliers. A further feature of our analysis is the treatment of purchases of capital goods. These contribute to labour cost savings by acting as both complements to some and substitutes for other workers; at the same time product quality improvements from business R&D are often embodied in new capital goods. We therefore treat capital goods as purchased inputs, analogous to intermediate goods, with purchases and sales between their sectors of origin and destination contributing to the transmission of labour productivity improvement and product innovation.

The focus of this chapter is thus on the use of domestic labour and on product improvement via own R&D. We abstract from the changes in the level of final demand in order to focus on the changing efficiency of supply. Elsewhere we have undertaken the full accounting of changes in output growth, including an assessment of the respective roles of domestic demand, technological change and the pattern of net trade (see Gregory and Greenhalgh, 1997) and we shall refer to these results where they bear on the present study.

Direct and indirect input–output labour intensity of sectoral outputs

Our first focus is on labour productivity and its transmission through the economy. In the input–output framework economy-wide labour use can be expressed as:

$$L = \ell'X = \ell'(I - A)^{-1}F \tag{1}$$

where L is total labour use (scalar), X and F are column vectors of gross output and final demand for domestic output by sector, $(I - A)^{-1}$ is the Leontief inverse matrix, and ℓ is the vector of labour requirements per unit of sectoral gross output. For each ith sector ℓ_i is therefore the inverse of labour productivity, measured as gross output per unit of labour input. The first line of equation (1) expresses total labour use in terms of sectoral gross outputs and their associated employment requirements. This focuses on the sector where the output and employment are located. However, much of sectoral output is sold on to other sectors before reaching its final use, a process which the input–output approach is designed to encapsulate. The second line of (1) uses the standard input–output relationship, $X = (I - A)^{-1}F$, to express total employment as a function of sectoral final demands for domestic goods and services. Final demand in sector i, F_i, gives rise to gross output in (all) other sectors, through the intermediate output requirements encapsulated in the Leontief inverse, and therefore to employment in (all) other sectors.

To capture this concept of the employment generated economy-wide to meet final demand in any sector we define 'input–output labour intensity' λ as:

$$\lambda' = \ell'(I - A)^{-1} \tag{2}$$

where ℓ will now be referred to as 'direct' labour productivity within each sector and $(I - A)^{-1}$ gives the inter-industry transmission of output requirements. Total employment in (1) can correspondingly be written more compactly as:

$$L = \lambda' F \tag{3}$$

Expanding this to the vector of input–output labour use across sectors gives:

$$\lambda' \hat{F} = \ell'(I - A)^{-1} \hat{F} \tag{4}$$

where the circumflex denotes the diagonal matrix formed from the vector. Each ith element of the employment vector on the left-hand side of (4) gives the labour use, economy-wide, required to produce sectoral final demand F_i; summing the elements of this vector gives the scalar L on the left-hand side of (3). Input–output labour intensity l in (2) evaluates economy-wide employment generated per unit of sectoral final demand, while labour use $\lambda' F$ in (4) scales this by the level of sectoral final demand to give the total level of employment generated.

Changes in labour intensity and labour use can be examined by differencing (2) and (4) over time. From (2):

$$\Delta \lambda' = \Delta \ell'(I - A)^{-1} + \ell'\{(I - A)^{-1}\} \tag{5}$$

This divides the change in labour intensity (labour cost-saving) between direct labour-saving $\Delta \ell$ and labour-saving through changes in the structure of inter-industry purchases, $\Delta(I - A)^{-1}$, for example through contracting-out.

The time-difference of (4) contains two components, a difference term in λ and one in $\hat{F}$. However, for the investigation of productivity change we abstract from changes in final demand and focus on the change in labour intensity, evaluated at actual levels of final demand:

$$\Delta \lambda' \hat{F} = \Delta \ell'(I - A)^{-1} \hat{F} + \ell'\{\Delta(I - A)^{-1}\} \hat{F} \tag{6}$$

The transmission of productivity change

We now extend the framework above to examine the transmission of labour cost-saving across sectors, distinguishing productivity gains originating with decisions by the sector itself from those derived through inter-industry purchases. This latter part, the transmission of productivity gains across sectors, or backward linkages, will be denoted 'input effects' (Postner and Wesa, 1983). It is also possible, and in terms of key sectors particularly illuminating, to look at this process from the other direction, following the sector's productivity-enhancing improvements forward in the supply chain, through its sales to further intermediate and final users ('forward' effects).

To examine the contribution of these various effects we decompose (5) and (6) in two ways. Using the expression:

$$(I - A)^{-1} = I + A + A^2 + A^3 + \ldots = I + A + \tilde{A} \tag{7}$$

and noting that $A = \hat{A} + (A - \hat{A})$ where $\hat{A}$ is the diagonal matrix formed from the principal diagonal of A, (5) can be expressed as:

$$\begin{aligned}\Delta\lambda' &= \Delta\ell'(I + A + \tilde{A}) + \ell'(\Delta A + \Delta\tilde{A}) \\ &= \Delta\ell'(I + \hat{A}) + \ell'(\Delta A) + \Delta\ell'(A - \hat{A} + \tilde{A}) + \ell'(\Delta\tilde{A})\end{aligned} \tag{8}$$

Combining the first and second pairs of terms, equation (8) can be summarised as:

$$\Delta\lambda' = \Delta\lambda'_{own} + \lambda'_{input} \tag{9}$$

where: $\Delta\lambda'_{own} = \Delta\ell'(I + \hat{A}) + \ell'\Delta A$

$\Delta\lambda'_{input} = \Delta\ell'(A - \hat{A} + \tilde{A}) + \ell'\Delta\tilde{A}$

This divides the change in total labour requirements per unit of final demand between 'own effects', $\Delta\lambda'_{own}$, and 'input effects', $\Delta\lambda'_{input}$. The 'own effects' comprise the change in the sector's direct labour use $\Delta\ell'$ for the production of its own output, $(I + \hat{A})$, and the labour requirements associated with changes in its own use of intermediate inputs, DA. The elements in the 'own effects' reflect decisions concerning factor proportions, the organisation of production, and intermediate purchases which are under the direct control of the firms in the sector. The 'input effects' comprise the change in labour requirements involved in the input use pattern of supplier industries $(A - \hat{A})$, plus the changes in labour requirements further back in the supply chain, encapsulated in $\tilde{A}$.

Applying the same decomposition to equation (6) gives analogous terms in 'own' and 'input' labour use

$$\Delta\lambda'\hat{F} = \Delta\lambda'_{own}\hat{F} + \Delta\lambda'_{input}\hat{F} \tag{10}$$

where $\Delta\lambda'_{own}\hat{F} = \Delta\ell'(I + \hat{A})\hat{F} + \ell'(\Delta A)\hat{F}$

and $\quad \Delta\lambda'_{input}\hat{F} = \Delta\ell'(A - \hat{A} + \tilde{A})\hat{F} + \ell'(\Delta\tilde{A})\hat{F}$

An equivalent, and perhaps more revealing, way of assessing the transmission of labour cost-saving through the economy is by following the supply chain in the opposite direction, forward from the industry originating the productivity change to the industries purchasing its output. This inverts the 'input effects' into 'forward effects':

$$\Delta\hat{\lambda}_{fwd}F = \Delta\hat{\ell}(A - \hat{A} + \tilde{A})F + \hat{\ell}(\Delta\tilde{A})F \tag{11}$$

Equation (11) evaluates the economy-wide labour-saving embodied in each sector's sales to intermediate users and to final demand. Implied in this is the fact that the demand-weighted sum across sectors of input use effects must be identically equal to the sum of all forward use effects. The parallel 'forward' effect for labour intensity, corresponding to a one-unit increase in final demand in all sectors simultaneously, gives little insight and will not be presented.

Endogenising investment

Purchases of new capital goods are classified on National Accounts conventions as an element of final demand, i.e. investment. Given that we want to trace the impact of labour productivity and product improvements through the forward and backward linkages it would be arbitrary and inappropriate to classify investment as a final demand. The more relevant perspective is that capital goods are analogous to intermediate goods, purchased from their sector of origin in order to contribute to further production. (A full discussion of the accounting principles in the input–output framework is given in Leontief, 1951, part I). Endogenising the demand for capital goods, the augmented input-output model in flow terms becomes:

$$X = AX + JX + C \tag{12}$$

where $J = [j_{ij}] = J_{ij}/X_j$ is the coefficient matrix of capital goods (GDFCF) purchases J_{ij} per unit of sectoral gross output, and C is the vector of final consumption (final demand less investment). In a dynamic Leontief framework J would represent requirements for maintaining a steady-state capital/output ratio given depreciation and the trend growth of output. Since our approach is purely a descriptive exercise in growth accounting, J represents actual purchases of capital goods. With capital goods purchases treated in this way, accounting consistency requires profits to be defined on a cash flow rather than a trading basis. Profits on this measure may become negative as borrowing occurs or assets are run down to finance the investment. Total purchases may then exceed the value of gross output.

From (12):

$$\mathbf{X} = (\mathbf{I} - \mathbf{A} - \mathbf{J})^{-1}\mathbf{C} = (\mathbf{I} - \mathbf{A}^*)^{-1}\mathbf{C} \tag{13}$$

where $\mathbf{A}^* = (\mathbf{A} + \mathbf{J})$ and the extended Leontief inverse $(\mathbf{I} - \mathbf{A}^*)^{-1}$ includes the coefficient matrix of capital purchases; gross output becomes a function of the consumption element of final demand only (private and public consumption of domestic output, plus exports).

With capital requirements endogenous, the change in input–output labour intensity (9) becomes:

$$\Delta\lambda' = \Delta\lambda'_{own} + \Delta\lambda'_{input} \tag{14}$$

where: $\Delta\lambda'_{own} = \Delta\ell'(\mathbf{I} + \hat{\mathbf{A}}^*) + \ell'\Delta\mathbf{A}^*$

and $\Delta\lambda'_{input} = \Delta\ell'(\mathbf{A}^* - \hat{\mathbf{A}}^* + \tilde{\mathbf{A}}^*) + \ell'\Delta\tilde{\mathbf{A}}^*$

Similarly, the change in input–output labour use (10) becomes:

$$\Delta\lambda'\hat{\mathbf{C}} = \Delta\lambda'_{own}\hat{\mathbf{C}} + \Delta\lambda'_{input}\hat{\mathbf{C}} \tag{15}$$

where: $\Delta\lambda'_{own}\hat{\mathbf{C}} = \Delta\ell'(\mathbf{I} + \hat{\mathbf{A}}^*)\hat{\mathbf{C}} + \ell'(\Delta\mathbf{A}^*)\hat{\mathbf{C}}$

and $\Delta\lambda'_{input}\hat{\mathbf{C}} = \Delta\ell'(\mathbf{A}^* - \hat{\mathbf{A}}^* + \tilde{\mathbf{A}}^*)\hat{\mathbf{C}} + \ell'(\Delta\tilde{\mathbf{A}}^*)\hat{\mathbf{C}}$

The 'forward' effects on labour use, corresponding to (11), are given by:

$$\Delta\lambda'_{fwd}\mathbf{C} = \Delta\hat{\ell}(\mathbf{A}^* - \hat{\mathbf{A}}^* + \tilde{\mathbf{A}}^*)\mathbf{C} + \hat{\ell}(\Delta\tilde{\mathbf{A}}^*)\mathbf{C} \tag{16}$$

(14), (15) and (16) will be the main expressions used in the empirical evaluation of the transmission of labour productivity change. While this deals explicitly with the possibility of capital–labour substitution in production, the method does not adjust for the possibility that some producers may be importing more labour intensive intermediate goods. We have evidence on this point from our other study (Gregory and Greenhalgh, 1997) and we shall comment on these research findings below.

R&D intensity and the transmission of improved product quality

A further major dimension of productivity growth is product innovation and enhanced product quality. The direct identification and measurement of changes in product quality poses many difficulties, but the framework developed above gives a way of examining the transmission of these changes across sectors. In what follows we posit that the flow of product improvements generated in any sector can be proxied by its level of current business R&D expenditure. By regarding these as embodied in the sector's outputs, of intermediate goods, new capital goods or commodities for final consumption, we trace their transmission through the economy via these market transactions (Griliches and Lichtenberg, 1984).

This transmission mechanism may also be regarded as proxying the spillover process in the dissemination of the improvements generated by R&D (Griliches, 1979). In the absence of an economy-wide set of hedonic price indices, the conventional methods of measuring real inter-industry purchases by deflating nominal output are unlikely to give full representation of quality improvements. For example, where unit price stays constant following product improvement, without hedonic adjustment the observed market prices of improved products will be too high relative to products of unchanged quality. The real volume of improving products may be persistently understated, in proportion to the extent of transactions. The analysis can also be viewed as tracing the social effects of R&D or perhaps even the 'knowledge spillover'. The full return to R&D is rarely captured by the spender, as the public good aspect of innovation brings imitation, the resulting competition forcing prices down, such that the purchaser pays less than the full user value of the embodied improvements. This additional social return can be assumed to be proportional to the value of transactions. The direction and scale of pure knowledge spillovers, the disembodied transfer of ideas across sectors, are difficult to infer on a systematic basis; if they are assumed to be proportional to the volume of market transactions between the sectors our transmission mechanism also covers this interpretation. Each of these effects is a plausible part of the diffusion process; our data do not allow us to distinguish among them.

The direct R&D content of sectoral gross output r is defined as R&D expenditure, R_i, per unit of sectoral gross output, $r_i = R_i/X_i$. The input–output (direct plus indirect) R&D intensity per unit of final consumption is then:

$$\rho' = r'(I - A^*)^{-1} \qquad (17)$$

This is the direct analogue of the input–output labour intensity λ above. We equate the rate of product quality improvement, Δq, with this R&D intensity of final consumption. This implies the assumption that sectoral R&D, and therefore product improvement, is embodied to an equal degree in the sector's output irrespective of purchaser. This is the standard 'product homogeneity' assumption of input–output models and parallels the method used by Scherer (1982, 1984) for patented inventions deemed to have a wide variety of general applications.

This measure of the change in embodied product quality can be decomposed as above into the amount attributable to the sector's own R&D (Δq_{own}) and the amount embodied in purchases from supplying sectors (Δq_{input}):

$$\Delta q' = \rho' = r'(I + \hat{A}^*) + r'(A^* - \hat{A}^* + \tilde{A}^*) = \Delta q'_{own} + \Delta q'_{input} \qquad (18)$$

where: $\Delta q'_{own} = r'(I + \hat{A}^*)$

$\Delta q'_{input} = r'(A^* - \hat{A}^* + \tilde{A}^*)$

Because the change in input product quality is observed within years there is no corresponding change in the A or J matrices; thus (18) is the within-year equivalent of (8) above.

Total R&D use embodied in each sector's final output can be derived by evaluating (18) at actual levels of final consumption:

$$\Delta q'\hat{C} = \Delta q'_{own}\hat{C} + \Delta q'_{input}\hat{C} \qquad (19)$$

where $\Delta q'_{own}\hat{C} = r'(I + \hat{A}^*)\hat{C}$

and $\quad \Delta q'_{input}\hat{C} = r'(A^* - \hat{A}^* + \tilde{A}^*)\hat{C}$

The 'forward linkages' transmission of R&D, which traces product quality improvements forward from the industry of origin, as in (16) above, is given by:

$$\Delta \hat{q}_{fwd}C = \hat{\rho}_{fwd}C = \hat{r}(A^* - \hat{A}^* + \tilde{A}^*)C \qquad (20)$$

Input–output data 1979–90

Input–output tables for the UK are available for 1979, 1985 and 1990 (BSO, 1983; CSO, 1989, 1994). All are based on the 1980 Standard Industrial Classification, and are in current prices only. The tables are constructed on a commodity by commodity basis. This is preferable to an industry by industry basis, as commodity-specific technologies are a more persuasive assumption than industry technologies. The level of commodity disaggregation has varied over time but is around 100 sectors, with a more detailed disaggregation for manufacturing than for services. A common 87-sector classification has been adopted, with outputs deflated to 1985 values by sectoral producers' output price indices. (Further details of these data issues and those discussed below are given in the Data Appendix to this chapter.)

Sales of output supplied as capital goods are available for all commodity sectors, but to purchasers classified by around 40 industries. Conversion from the commodity by industry to a commodity by commodity basis has been carried out using the 'make' matrix for the appropriate year. Under the assumption that homogeneous sectoral outputs are sold to all purchasers, sales of capital goods have been deflated by the producers' price index as applied to sectoral gross output.

For constructing our measure of sectoral labour demand we use employment income generated in the production of gross output, taken directly from the value added entries in the input–output tables. This was converted to real terms by revaluation at constant 1985 hourly earnings (see Appendix for details). This measures real outlay on labour from the employer's perspective, in effect combining a head-count of workers and hours, weighted by earnings levels, so avoiding the need to control for changes in proportions of full- and part-time workers. Furthermore any substitution of a skilled worker for one or more less-skilled workers which causes a rise in employment income is reflected as a rise in labour demand, following the concept of efficiency units of labour used in the growth literature. A further advantage of this measure is that the input–output framework imposes consistency between the measure of labour cost on the input side and sectoral outputs, including an exact match in sectoral classification. However there is an unavoidable problem arising from rising self-employment during the period of this study as the incomes of the self-employed are classified to profits. A switch in employment status from employed to self-employed creates an inflation of profit compared with employment income and may cause a small upward bias in the assessment of increased labour efficiency.

Our measure of R&D is UK business expenditure on R&D (BERD) (CSO, 1996). Unfortunately, while these data are supplied for 33 sectors they are on the 1992 Standard Industrial Classification. Aligning with the 1980 SIC reduces the available disaggregation to 19 sectors. In recognition of the fact that R&D expenditure has a very different composition from sectoral outputs or inputs, deflation to constant 1985 prices has been carried out using the deflators developed by Cameron (1996).

An immediate issue in the empirical implementation of the model is the form in which R&D expenditure best represents our variable of interest, the rate of product innovation or product quality improvement (Griliches, 1995). We use current-dated R&D expenditure. Several points can be made in support of this. Business R&D is often concerned with near-market research to refine prototypes and commercialise developments from more fundamental research; this indicates relatively short lags. It also suggests that the rate of depreciation – a problematic issue in the construction of a 'stock' measure of business R&D – may be quite high. Moreover, although levels of R&D differ widely and persistently across sectors, sectoral R&D intensity changes only slowly; any potential distributed-lag or cumulated stock series would show only limited variability over time. Arguments similar to some of these underpin the extensive use of the Terleckyj transformation (Terleckyj, 1974, 1980). As a further practical consideration, R&D data are not available prior to our data period at the level of disaggregation which we use; the construction of a distributed-lag variable would therefore involve the sacrifice of current sample information. We are supported by other studies which have found that R&D stock and flow measures are equally useful in panel data analysis (Hall, 1993 and references there). In practice we faced the further limitation that industry-level R&D data are available only at a more aggregated level before 1981. However, we found that the 1981 totals were closely similar to those for 1979, having peaked in 1980, and thus we adopted the 1981 figures to gain the necessary industry breakdown.

The analysis is carried out for three dates 1979, 1985 and 1990, determined by the availability of input–output data. The first and last dates were around cyclical peaks, indicating that capacity utilisation should be reasonably comparable between them. With the 1980s in effect spanning one long cycle, 1985 was a year of recovery but neither a peak nor trough. In spite of this asymmetry, the two sub-periods can offer a useful contrast of periods of recession and recovery.

A 15-sector level of aggregation has been applied in all the computations. This is close to the maximum level of disaggregation consistently attainable, as limited by the R&D data. The only further aggregation, of several smaller manufacturing sectors, reducing the number of sectors from 19 to 15, aims to keep sectors at roughly equal size and maintain balance between the number of manufacturing sectors and those in service activities. Conditions for consistent aggregation have been extensively discussed in the input–output context, as aggregation across unlike activities introduces bias into estimates of coefficients. The extent of the bias is an empirical issue, depending on the characteristics of the data in each case. For the period we analyse it appears to be minimal.

The most extreme test for aggregation bias which we can apply is based on labour cost-saving without endogenous capital, for which data are available at 87 sectors. Since the focus of our analysis is the decomposition of changes over time we applied this approach to the first part of equation (1), decomposing the change in employment between 1979 and 1990 between the change in output and the change in direct labour-intensity:

Table 3.1 *Comparison of results calculated at 87- and 15-sector level (% change)*

	1979–90		1979–85		1985–90	
	Output	Labour intensity	Output	Labour intensity	Output	Labour intensity
	$\ell'\Delta X$	$\Delta\ell'X$	$\ell'\Delta X$	$\Delta\ell'X$	$\ell'\Delta X$	$\Delta\ell'X$
87 sectors	34.2	–39.6	9.2	–20.3	23.6	–17.2
15 sectors	33.7	–39.1	9.0	–20.1	23.5	–17.2

$$\Delta L = \ell'\Delta X + \Delta\ell'X \tag{21}$$

This was computed at 87 sectors and the estimates aggregated to 15 sectors. For comparison the 87-sector levels of output and employment were pre-aggregated to 15 sectors, the new Leontief inverse computed, and the decomposition repeated. As table 3.1 shows, the results were virtually identical.

Results I: Labour productivity growth and transmission

The main vehicle for our analysis of labour productivity change is the sectoral input–output (I–O) labour intensity λ, the labour required through all lines of business, including supplies of capital goods, to generate a unit of the sector's output, a commodity for final consumption. As defined in equation (2) this depends on the structure of the sector's inter-industry demands, encapsulated in the extended Leontief inverse $(I - A^*)^{-1}$, in conjunction with direct labour intensity ℓ in each sector. It can usefully be thought of as a weighted sum of direct labour intensities in all sectors, with the weights determined by this structure of sectoral input purchases.

Direct labour-intensity is shown in table 3.2 columns 1–3; the range across sectors is wide, in 1979 from 75p of labour cost per £1 of gross output in personal and public services to just 13p in energy and utilities, and from 56p to 8p in 1990. The conventional view that services are more labour intensive than manufacturing is clearly confirmed. (Manufacturing is defined as sectors 3–10, services as 12–15.) When labour intensity is measured on the I–O basis λ, shown in table 3.2 columns 4–6, the levels are uniformly higher, partly because these are being expressed per unit of the sector's final consumption rather than its gross output. The range narrows substantially, in 1990 from 7:1 to under 3:1, reflecting the

Table 3.2 *Levels of direct and I–O labour intensities*

	Direct labour intensity ℓ £ employment income per £ gross output			Indirect labour intensity λ £ employment income per £ final consumption			Sectoral Leontief multiplier Column sum of $(I - A^*)^{-1}$		
	1979	1985	1990	1979	1985	1990	1979	1985	1990
1 Agriculture	0.18	0.11	0.12	0.72	0.48	0.37	2.78	2.55	2.20
2 Energy and utilities	0.13	0.11	0.08	0.37	0.34	0.26	1.83	1.95	2.07
3 Food & drink	0.16	0.15	0.14	0.62	0.52	0.42	2.50	2.55	2.36
4 Paper & printing	0.35	0.33	0.25	0.71	0.61	0.49	1.91	1.90	1.93
5 Metals	0.32	0.24	0.22	0.77	0.53	0.46	2.31	2.04	1.99
6 Manufacturing nes	0.31	0.29	0.26	0.65	0.54	0.48	2.00	1.87	1.91
7 Mechanical engineering	0.37	0.35	0.29	0.76	0.67	0.56	2.03	2.08	2.07
8 Chemicals	0.21	0.17	0.15	0.64	0.43	0.38	2.35	2.05	1.97
9 Electrical equipment	0.43	0.31	0.26	0.79	0.58	0.51	1.96	1.93	1.96
10 Transport equipment	0.41	0.32	0.23	0.81	0.62	0.46	2.13	2.01	1.91
11 Construction	0.32	0.22	0.18	0.72	0.55	0.51	2.20	2.17	2.43
12 Trade & catering	0.41	0.36	0.34	0.72	0.64	0.61	1.82	1.95	2.11
13 Transport & comm.	0.43	0.40	0.32	0.68	0.64	0.58	1.69	1.84	2.04
14 Business services	0.63	0.35	0.28	1.30	0.72	0.73	2.80	2.24	2.85
15 Personal & public servs	0.75	0.64	0.56	0.86	0.74	0.75	1.28	1.34	1.75
Manufacturing	0.30	0.26	0.22	0.70	0.55	0.46	2.20	2.12	2.06
Services	0.56	0.45	0.38	0.84	0.69	0.68	1.68	1.71	2.02
Total	0.39	0.33	0.28	0.75	0.61	0.57	1.87	1.87	2.05

'weighted sum' structure. (The value of λ of £1.30/£ in business services in 1979 is the only instance where capital goods purchases in an investment boom are sufficiently large relative to trading profits for input purchases to exceed gross output; see the discussion on pages 64–5 above.) Individual sector rankings remain broadly similar, with the correlation coefficient between 0.8 and 0.9 in each year. Services are again more labour-using than manufacturing, but by a reduced margin, as manufacturing tended in the past to use more intermediate inputs, so drawing on more employment through the supply chain; however as can be seen the two broad aggregates have drawn closer together through time (columns 7–9).

On both measures the labour-intensity of production has been falling in all sectors of the economy over the 1980s; the aggregate effect of technical and structural change has been labour-saving almost universally. Perhaps contrary to the received view, direct labour-saving has proceeded more rapidly, both absolutely and proportionally, in services than in manufacturing. These aggregates, however, span considerable diversity of performance. The three top performers comprise one sector in services and two in manufacturing (business services, transport equipment and electrical equipment) as do the three poorest performers (trade & catering, food & drink, and manufacturing nes). On the I–O measure the pattern is much more clearcut. Manufacturing moves clearly ahead of services, with annual productivity gains of 3.8 per cent, double the rate for services. This superiority of manufacturing is broadly based, with all sectors out-performing each of the services sectors apart from business services. The comparative record of these two measures of labour productivity growth for Britain in the 1980s indicates that the traditional view of manufacturing as the superior source of productivity gains (Rowthorn and Wells, 1987, Chapter 1) has to be formulated with care. *It is in the production of manufactures, rather than in manufacturing establishments, that the labour productivity performance has been notable; but the production of business services ranks higher than any manufacturing supply chain in its overall productivity performance.*

Alongside direct productivity change the important further element incorporated into the change in I–O productivity $\Delta\lambda$ is structural change in the use of inputs, measured through changes in the Leontief inverse. This is shown in table 3.3. Two features are of key importance. The first is the greatly increased use of business services for every type of output. This was evident for services at least as strongly as for manufactures, with the share of business services in production costs typically rising by between 10 and 20 percentage points. Part of this is likely to reflect the increased service

Table 3.3 The change in the Leontief inverse 1979–90
$\Delta[(I-A^*)^{-1}]$

Sectors	1	2	3	4	5	6	7	8	9	10	11	12	13	14	15
1 Agriculture	-0.076	0.001	-0.031	0.001	-0.001	0.004	0.000	-0.005	0.001	0.000	0.001	0.005	0.003	0.002	0.003
2 Energy & utilities	-0.077	0.236	-0.038	-0.028	-0.113	-0.062	-0.029	-0.120	-0.045	-0.078	-0.027	-0.003	-0.005	-0.017	0.016
3 Food & drink	-0.126	0.002	-0.046	0.000	-0.002	0.003	0.000	-0.016	0.002	-0.001	0.004	0.011	0.010	0.007	0.008
4 Paper & printing	-0.020	0.001	-0.021	-0.078	-0.017	-0.012	-0.003	-0.018	0.000	-0.007	-0.001	-0.005	-0.014	-0.017	0.005
5 Metals	-0.027	-0.018	-0.009	-0.006	-0.050	-0.017	-0.022	-0.037	-0.059	-0.087	-0.017	-0.001	-0.006	-0.040	0.009
6 Mfg nes	-0.032	-0.018	0.001	-0.002	-0.026	-0.057	-0.017	-0.016	-0.004	-0.012	-0.046	-0.003	0.004	-0.050	0.018
7 Mech. eng.	-0.052	0.005	-0.025	0.002	-0.036	-0.018	0.036	-0.038	-0.017	-0.047	-0.001	-0.002	0.001	-0.025	0.008
8 Chemicals	-0.032	-0.008	-0.019	-0.017	-0.020	-0.041	-0.012	-0.154	-0.015	-0.024	-0.022	-0.006	-0.004	-0.024	0.003
9 Elect. eqpt.	-0.003	0.011	0.001	0.002	-0.003	0.003	-0.004	-0.003	-0.001	-0.009	0.003	0.006	0.022	-0.013	0.010
10 Transp. eqpt.	-0.013	0.005	-0.007	-0.001	-0.012	-0.005	0.002	-0.005	0.001	0.012	-0.002	-0.006	-0.006	-0.046	0.009
11 Construction	-0.023	-0.040	0.026	0.053	0.004	0.025	0.032	0.014	0.039	0.020	0.118	0.064	0.067	-0.092	0.119
12 Trade & catering	-0.097	-0.012	-0.051	-0.046	-0.100	-0.018	-0.046	-0.028	-0.012	-0.062	-0.028	-0.010	0.011	-0.054	0.007
13 Transport & comm.	-0.031	0.000	-0.027	-0.013	-0.039	-0.004	-0.020	-0.032	-0.015	-0.033	-0.011	0.036	0.070	-0.048	0.019
14 Business services	0.124	0.089	0.175	0.215	0.135	0.154	0.161	0.160	0.184	0.144	0.269	0.227	0.196	0.507	0.194
15 Personal & public servs	-0.091	-0.013	-0.063	-0.063	-0.043	-0.041	-0.046	-0.079	-0.048	-0.037	-0.014	-0.026	-0.004	-0.038	0.041
Total	-0.575	0.240	-0.134	0.021	-0.322	-0.086	0.032	-0.375	0.010	-0.221	0.227	0.286	0.345	0.051	0.470

Note: The ijth element gives the change in sector j's purchases, direct and indirect, of intermediate and capital goods from sector i, to meet one unit of final consumption in sector j. The column sum therefore expresses the change in total inputs used to meet one unit of final consumption in sector j.

content of commodity supply, for example through marketing, software development and information technology support. A further part will represent the contracting-out of conventional service functions, such as recruitment, tax and accountancy, to consultancies and specialist agencies. This parallels the shift towards transactional services and out-sourcing in the US economy described by Blair and Wyckoff (1989). The second major development has been the trend towards increased use of inputs in services production, notably in personal and public services (some elements of which were affected by privatisation), against a trend towards input-saving for manufactures. Services have increased their input purchases, for example from transport & communications, as well as on the investment side from construction and electrical equipment. For manufactures, although the changes in the use of intermediates are individually quite small, their cumulative effect has been to generate significant input-saving in certain sectors. *To sum up, the phenomenon of 'contracting-out' of services supply was quite general, not confined to manufacturing or to services affected by privatisation; inter-industry purchases mostly fell between manufacturers and rose between service providers.*

Following this direct look at the changing Leontief inverse, for each commodity the total labour-saving on the I–O measure can be attributed, using equation (5) above, between direct labour saving in all supplying sectors $\Delta \ell$, and changes in the structure of inter-industry purchases, $\Delta(I - A^*)^{-1}$. Table 3.4 presents these results for 1979–90. The message can be summarised simply: *direct labour use declined in the production of all commodities, but the changes in inter-industry purchases for the production of services were job-creating, while those for the production of manufactures were generally job-destroying.* The same broad picture emerges when this breakdown is evaluated at average final consumption over the period, reflecting the impact of the differing sizes of sectors (table 3.4, lower panel, and figure 3.1). However, size matters for total impact: although labour-saving was greater for manufactures, at 24p/£ against 16p for services, the larger size of the services' final demands meant that labour-saving in their supply chains at £30 billion reduced total employment by more than for manufactures (£25 billion). In the production of services and construction the increased use of intermediates generated employment growth in supplying sectors which partially offset direct job-shedding. In the production of manufactures, on the other hand, the trend to labour-saving tended to be reinforced by input-saving more widely.

How far might these results have been affected by changing patterns of intermediate imports? In Gregory and Greenhalgh (1997, table 10) we showed that the employment effects of import penetration during this

Table 3.4 *Change in labour intensity and labour use 1979–90*

£ employment income per £ final consumption	Total $\Delta\lambda$	Labour intensity Labour saving $\Delta\ell'(I-A^*)^{-1}$	I–O change $\ell'\Delta((I-A^*)^{-1})$
1 Agriculture	−0.35	−0.21	−0.15
2 Energy & utilities	−0.12	−0.15	0.04
3 Food & drink	−0.20	−0.17	−0.02
4 Paper & printing	−0.22	−0.24	0.02
5 Metals	−0.31	−0.24	−0.08
6 Manufacturing nes	−0.17	−0.17	0.00
7 Mechanical engineering	−0.20	−0.22	0.02
8 Chemicals	−0.27	−0.19	−0.08
9 Electrical equipment	−0.28	−0.30	0.02
10 Transport equipment	−0.35	−0.31	−0.05
11 Construction	−0.21	−0.32	0.10
12 Trade & catering	−0.10	−0.21	0.11
13 Transport & comms	−0.10	−0.24	0.13
14 Business services	−0.57	−0.65	0.08
15 Personal & public servs	−0.11	−0.29	0.18
Manufacturing	−0.24	−0.22	−0.02
Services	−0.16	−0.29	0.14
Total	−0.18	−0.26	0.07

£ million	$\Delta\lambda\hat{F}$	Labour use $\Delta\ell'(I-A^*)^{-1}\hat{F}$	$\ell'\Delta((I-A^*)^{-1})\hat{F}$
1 Agriculture	−1481	−873	−608
2 Energy & utilities	−3437	−4579	1142
3 Food & drink	−5073	−4499	−574
4 Paper & printing	−1088	−1172	84
5 Metals	−1823	−1380	−443
6 Manufacturing nes	−2950	−2894	−56
7 Mechanical engineering	−2051	−2259	208
8 Chemicals	−3420	−2436	−984
9 Electrical equipment	−3234	−3484	250
10 Transport equipment	−5022	−4369	−652
11 Construction	−563	−837	273
12 Trade & catering	−7104	−14468	7365
13 Transport & comms	−2063	−4757	2694
14 Business services	−11309	−12930	1622
15 Personal & public servs	−9069	−23327	14258
Manufacturing	−24662	−22494	−2168
Services	−30108	−56320	26212
Total	−59688	−84266	24578

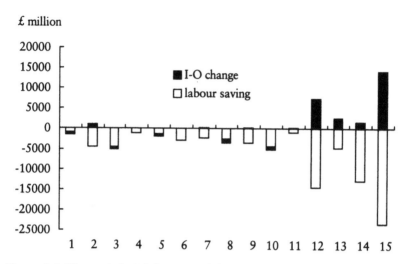

Figure 3.1 *Change in I–O labour use; labour saving vs I–O change 1979–90*
Key: 1 Agriculture; 2 Energy and utilities; 3 Food and drink; 4 Paper and printing; 5 Metals; 6 Manufacturing nes; 7 Mechanical engineering; 8 Chemicals; 9 Electrical equipment; 10 Transport equipment; 11 Construction; 12 Trade and catering; 13 Transport and communications; 14 Business services; 15 Personal and public services.

period largely arose from the loss of final goods markets; the effects of rising purchases of intermediates were very muted. Added to this, the differentials between manufactures and services in their use of intermediate inputs would act against the finding of higher productivity in the supply chain for business services, so this result is robust. In table 3.4 we see that production of high technology manufactures (chemicals, transport equipment and electrical equipment) achieved among the highest efficiency gains, as measured by I–O labour intensity. These productivity gains in high tech products would support the strong export performance of high technology manufactures noted in our earlier work (Gregory and Greenhalgh, 1997, tables 3, 6). However, because of the smaller size of these sectors, the overall efficiency gains measured by total labour saving attributable to each of them are substantially less than for personal & public services in spite of its low productivity growth.

Looking briefly at these developments over the sub-periods 1979–85 and 1985–90, table 3.5 shows a clear sequencing. In the early 1980s recessionary period (1979–85), direct labour saving dominates, with labour cost being reduced by an average of 14p/£, while changes in

Table 3.5 *Change in I–O labour intensities 1979–85 and 1985–90*

£ employment income per £ final consumption	1979–85 Total $\Delta\lambda$	1979–85 Labour saving $\Delta\ell'(I - A^*)^{-1}$	1979–85 I–O change $\ell'\Delta((I - A^*)^{-1})$	1985–90 Total	1985–90 Labour saving	1985–90 I–O change
1 Agriculture	−0.25	−0.18	−0.07	−0.11	−0.04	−0.07
2 Energy & utilities	−0.03	−0.08	0.04	−0.08	−0.08	0.00
3 Food & drink	−0.10	−0.11	0.01	−0.10	−0.07	−0.03
4 Paper & printing	−0.11	−0.09	−0.01	−0.11	−0.13	0.02
5 Metals	−0.24	−0.16	−0.08	−0.08	−0.07	−0.01
6 Manufacturing nes	−0.11	−0.08	−0.04	−0.05	−0.08	0.02
7 Mechanical engineering	−0.09	−0.11	0.02	−0.11	−0.11	0.00
8 Chemicals	−0.21	−0.12	−0.09	−0.06	−0.06	0.01
9 Electrical equipment	−0.21	−0.21	0.00	−0.07	−0.09	0.02
10 Transport equipment	−0.19	−0.17	−0.03	−0.16	−0.14	−0.02
11 Construction	−0.18	−0.19	0.02	−0.04	−0.11	0.07
12 Trade & catering	−0.08	−0.13	0.05	−0.03	−0.08	0.05
13 Transport & comm.	−0.04	−0.09	0.04	−0.06	−0.13	0.07
14 Business services	−0.59	−0.42	−0.16	0.02	−0.16	0.17
15 Personal and public servs	−0.12	−0.15	0.02	0.01	−0.11	0.12
Manufacturing	−0.15	−0.13	−0.02	−0.09	−0.09	0.00
Services	−0.15	−0.16	0.01	−0.01	−0.11	0.10
Total	−0.14	−0.14	0.00	−0.04	−0.10	0.05

inter-industry purchases were neutral overall in their impact. In 1985–90, on the other hand, labour saving continued, but at a reduced rate. However, the increased purchases by services and of services began to have a positive impact. In this respect the second half of the 1980s marks a significant phase in the development of the UK as a services-based economy.

In the above discussion note that in terms of I–O productivity change the output of each sector benefits from productivity growth in all sectors. Following equation (9) we now divide the changes into 'own' effects, attributable to decisions made by the sector itself, and 'input' effects embodying direct labour saving by the sectors from which inputs have been purchased, and further back in the supply chain. Table 3.6 (cols 1–4) shows the input changes subdivided in this way. This four-way decomposition demonstrates conclusively that the biggest productivity gains were made as a result of 'own' within-sector effects (col. 1). The direct rationalisation of labour purchases, totalling £44 billion, comprised 74 per cent of total labour saving. These reductions occurred across all sectors, with the biggest savings being made in services, in particular personal & public services and business services. Substantial economies were also made as input purchases further back in the supply chain embodied fewer labour resources (col. 3); once again direct labour saving, now embodied in input purchases, had a larger impact on labour use than business reorganisation (col. 4). Services in particular benefited from buying in efficiency gains made in other sectors, with trade & catering the largest beneficiary, acquiring an estimated £9 billion of labour saving embodied in inputs such as food & drink, and transport & communications. For food & drink production these bought-in effects dominate all others, contributing 75 per cent of the overall £5 billion fall in labour use.

Forward linkages give an alternative perspective on the role of individual sectors in economy-wide productivity change, through the sales of intermediate goods and services to the rest of the economy. These forward linkages by sector, reflecting equation (11) above, are also presented in table 3.6, (cols 5 and 6). The dramatic implication to emerge, which is also pictured in figure 3.2, is the dominant role of the business services sector in the forward supply chain. It dominates in two quite separate respects. The first is labour saving/productivity gains; of the total £40 billion in labour saving transferred across sectors through input purchases/sales, business services supplies £16 billion (table 3.6 col. 5). However, these huge labour-saving effects which it has generated internally and transferred across other sectors have been counterbalanced by the expanded demand for business services from other sectors of £18 billion (table 3.6 col. 6). Personal & public services, by contrast, which recorded major within-sec-

Table 3.6 *Change in I–O labour use 1979–90: backward and forward linkages*

£ million	Own labour saving (a)	Own I–O change (b)	Input labour saving (c)	Input I–O change (d)	Lab. saving supplied (e)	Supplier (I–O) change (f)	Total input	Total supplied
1 Agriculture	−288	−287	−585	−321	−617	−56	−1481	−1249
2 Energy and utilities	−2167	50	−2412	1092	−1550	49	−3437	−3618
3 Food & drink	−691	−302	−3808	−273	−203	−78	−5073	−1274
4 Paper & printing	−575	−117	−597	201	−1215	−75	−1088	−1982
5 Metals	−683	−285	−697	−158	−1674	−772	−1823	−3415
6 Manufacturing nes	−1010	−279	−1884	223	−1031	−264	−2950	−2584
7 Mechanical engineering	−969	−74	−1290	283	−1050	−312	−2051	−2406
8 Chemicals	−816	−613	−1620	−370	−410	−564	−3420	−2404
9 Electrical equipment	−2104	−97	−1380	347	−1517	449	−3234	−3269
10 Transport equipment	−2651	−436	−1718	−216	−1440	11	−5022	−4516
11 Construction	−462	64	−375	209	−6552	3792	−563	−3158
12 Trade & catering	−4971	1930	−9497	5435	−1334	−1265	−7104	−5641
13 Transport & comm.	−2364	1023	−2393	1672	−2929	851	−2063	−3420
14 Business services	−8366	−109	−4564	1730	−16229	18369	−11309	−6335
15 Personal and public servs	−16226	6273	−7101	7985	−2171	−2295	−9069	−14419
Manufacturing	−9500	−2205	−12995	37	−8539	−1605	−24662	−21849
Services	−31928	9116	−23555	16822	−22662	15659	−29545	−29815
Total	−44344	6739	−39921	17840	−39921	17840	−59688	−59688

Notes: (a) $\Delta \ell'(I+\hat{A}^*)\hat{C}$; (b) $\ell'(\Delta A^*)\hat{C}$; (c) $\Delta \ell'(A^* - \hat{A}^* + \tilde{A}^*)\hat{C}$; (d) $\ell'\Delta \tilde{A}^*\hat{C}$; (e) $\Delta \hat{\ell}(A^* - \hat{A}^* + \tilde{A}^*)C$; (f) $\hat{\ell}\Delta \tilde{A}^*C$.

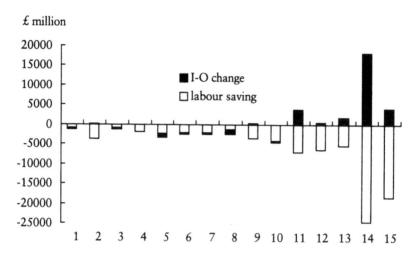

Figure 3.2 *Change in I–O labour use; forward linkages labour saving vs I–O change 1979–90*
Key: 1 Agriculture; 2 Energy and utilities; 3 Food and drink; 4 Paper and printing; 5 Metals; 6 Manufacturing nes; 7 Mechanical engineering; 8 Chemicals; 9 Electrical equipment; 10 Transport equipment; 11 Construction; 12 Trade and catering; 13 Transport and communications; 14 Business services; 15 Personal and public services.

tor productivity gains of £16bn, do not figure as strongly in forward productivity transmission, being mostly used in final demand.

Results II: Product quality growth and transmission

Turning to the sources and transmission of R&D, Table 3.7 shows the direct R&D intensities of gross output by sector, r above. This displays the well known concentration of British R&D in a narrow band of the manufacturing sector, namely chemicals, electrical equipment and transport equipment. All of these sectors undertook between 4p and 7p of R&D per £ of gross output in 1990. Outside these sectors and apart from mechanical equipment and business services, very little R&D is reported as being done, the intensity of R&D spending being under 1p/£. Moreover, the R&D intensity of output fell during the 1980s in all these major contributors to product quality improvement, with the exception of chemicals. In the electrical equipment sector R&D intensity fell from 7.2p/£ to 5.7p/£, while in transport equipment it fell from 6.3p/£ to 4.7p/£. Meanwhile, in chemicals it increased from 3.7p/£ to 6.6p/£.

However, when R&D intensities are measured on the I–O basis ρ, including the indirect content of R&D bought in from other sectors (equation 17) this picture is significantly modified (table 3.7, lower panel). *The spreading effects of inter-industry purchases of intermediates and capital goods ensure that at least some domestic R&D content is effectively present in either the new process technology or the improved final products across the whole range of goods and services delivered to consumers.*

Table 3.8 partitions the I–O R&D intensity into the part generated within the sector and the part acquired through purchases of intermediate or capital goods from the R&D spenders (equation 18). For the big R&D spenders 'own' R&D dominates, providing at least 80 per cent, sometimes 90 per cent of the total used. Among low technology manufacturers (sectors 3–7) the proportion of 'own' R&D falls to just 30 per cent. In this respect they are more similar to services than to the high technology sectors. Trade & catering presents a limiting case, with almost 100 per cent of the R&D embodied in production being bought in from outside.

Turning to the levels of R&D use, table 3.9 presents evidence of both backward and forward linkages after scaling for the size of sectoral final consumption, including the split between R&D undertaken within sector and that purchased or supplied (equation 19). As we found for productivity growth, the large and increasing size of the service sectors converts the low levels of R&D intensity (from table 3.8) into substantial, and increasing, inter-industry flows of prospects for enhanced product quality. Trade & catering again provides the most striking example. Its direct R&D expenditure has been negligible; on the I–O measure its R&D intensity remains one of the lowest; but when scaled by the size of final consumption, its R&D use places it behind only the 'major three' in 1985 and 1990. R&D use in the trade & catering sector increased by 44 per cent over the 1980s, to £615 million, against £953 billion in transport equipment, the smallest of the major three. Personal & public services provide a similar example of all these phenomena. *The I–O perspective on the transmission of R&D embodied in inter-industry product sales thus shows the service sectors as major users of R&D, even where their direct R&D spending is negligible.*

Returning to the theme of key sectors, we see that the principal generators of R&D are the same at the start and the end of the period. The major three – chemicals, electrical equipment and transport equipment – dominate. The remaining figures in table 3.9 give the forward impact of R&D from each sector through its inter-industry sales (equation 20). *Whilst the high technology manufacturing undertakers of R&D are also the major*

Table 3.7 *Direct and indirect R&D intensities*

£ R&D per £ gross output	Direct R&D intensity r		
	1979	1985	1990
1 Agriculture	0.001	0.001	0.003
2 Energy & utilities	0.004	0.004	0.004
3 Food & drink	0.003	0.003	0.004
4 Paper & printing	0.001	0.001	0.002
5 Metals	0.005	0.005	0.004
6 Manufacturing nes	0.003	0.003	0.003
7 Mechanical engineering	0.012	0.012	0.010
8 Chemicals	0.037	0.045	0.066
9 Electrical equipment	0.072	0.061	0.057
10 Transport equipment	0.063	0.058	0.047
11 Construction	0.000	0.000	0.000
12 Trade & catering	0.000	0.000	0.000
13 Transport & comm.	0.003	0.004	0.003
14 Business services	0.015	0.007	0.007
15 Personal & public servs	0.000	0.000	0.000
Manufacturing	0.011	0.023	0.024
Services	0.002	0.001	0.001
Total	0.009	0.008	0.009

£ R&D per £ final consumption	I-O R&D intensity ρ		
	1979	1985	1990
1 Agriculture	0.013	0.013	0.014
2 Energy & utilities	0.010	0.012	0.012
3 Food & drink	0.012	0.013	0.012
4 Paper & printing	0.008	0.008	0.009
5 Metals	0.015	0.012	0.012
6 Manufacturing nes	0.012	0.012	0.011
7 Mechanical engineering	0.022	0.023	0.020
8 Chemicals	0.054	0.058	0.080
9 Electrical equipment	0.085	0.075	0.069
10 Transport equipment	0.078	0.071	0.059
11 Construction	0.009	0.008	0.009
12 Trade & catering	0.007	0.008	0.008
13 Transport & comm.	0.013	0.014	0.014
14 Business services	0.031	0.016	0.019
15 Personal & public servs	0.002	0.003	0.005
Manufacturing	0.017	0.034	0.034
Services	0.008	0.007	0.008
Total	0.017	0.016	0.017

Table 3.8 *I–O R&D intensities: backward linkages*

£ R&D per £ final consumption	1979 Own	1979 Input	1985 Own	1985 Input	1990 Own	1990 Input
1 Agriculture	0.001	0.012	0.001	0.012	0.003	0.010
2 Energy & utilities	0.005	0.005	0.005	0.007	0.006	0.006
3 Food & drink	0.004	0.008	0.004	0.009	0.004	0.008
4 Paper & printing	0.002	0.006	0.002	0.006	0.002	0.007
5 Metals	0.006	0.009	0.006	0.006	0.005	0.007
6 Manufacturing nes	0.003	0.009	0.004	0.008	0.003	0.008
7 Mechanical engineering	0.013	0.009	0.014	0.009	0.012	0.008
8 Chemicals	0.045	0.009	0.051	0.008	0.073	0.007
9 Electrical equipment	0.079	0.007	0.069	0.007	0.062	0.007
10 Transport equipment	0.067	0.011	0.061	0.010	0.051	0.008
11 Construction	0.001	0.008	0.001	0.007	0.000	0.008
12 Trade & catering	0.000	0.008	0.000	0.008	0.000	0.008
13 Transport & comm.	0.003	0.010	0.004	0.010	0.004	0.010
14 Business services	0.016	0.015	0.008	0.008	0.009	0.010
15 Personal & public servs	0.000	0.002	0.000	0.003	0.000	0.005
Manufacturing	0.027	0.009	0.026	0.008	0.026	0.008
Services	0.002	0.006	0.001	0.006	0.001	0.007
Total	0.010	0.007	0.009	0.007	0.010	0.007
	Proportion Own	Proportion Input	Proportion Own	Proportion Input	Proportion Own	Proportion Input
1 Agriculture	0.08	0.92	0.08	0.92	0.25	0.75
2 Energy & utilities	0.48	0.52	0.41	0.59	0.48	0.52
3 Food & drink	0.31	0.69	0.31	0.69	0.32	0.68
4 Paper & printing	0.21	0.79	0.20	0.80	0.22	0.78
5 Metals	0.39	0.61	0.47	0.53	0.43	0.57
6 Manufacturing nes	0.29	0.71	0.33	0.67	0.26	0.74
7 Mechanical engineering	0.60	0.40	0.61	0.39	0.59	0.41
8 Chemicals	0.83	0.17	0.87	0.13	0.92	0.08
9 Electrical equipment	0.92	0.08	0.91	0.09	0.90	0.10
10 Transport equipment	0.86	0.14	0.86	0.14	0.86	0.14
11 Construction	0.07	0.93	0.08	0.92	0.03	0.97
12 Trade & catering	0.00	1.00	0.00	1.00	0.00	1.00
13 Transport & comm.	0.25	0.75	0.28	0.72	0.27	0.73
14 Business services	0.51	0.49	0.49	0.51	0.48	0.52
15 Personal & public servs	0.05	0.95	0.04	0.96	0.02	0.98
Manufacturing	0.76	0.24	0.76	0.24	0.77	0.23
Services	0.25	0.75	0.18	0.82	0.16	0.84
Total	0.59	0.41	0.58	0.42	0.57	0.43

Table 3.9 I–O R&D use: backward and forward linkages

£ million	1979 Own	1979 Input	1979 Supplied	1985 Own	1985 Input	1985 Supplied	1990 Own	1990 Input	1990 Supplied
1 Agriculture	3	30	9	4	48	10	20	60	30
2 Energy & utilities	131	142	97	165	237	98	182	200	124
3 Food & drink	96	221	23	96	218	27	101	209	35
4 Paper & printing	7	27	16	7	27	14	11	38	21
5 Metals	34	53	84	29	34	66	30	40	72
6 Manufacturing nes	60	152	55	57	114	62	47	134	57
7 Mechanical engineering	151	101	149	121	77	138	112	76	138
8 Chemicals	468	98	318	667	100	274	1121	102	408
9 Electrical equipment	735	64	538	879	84	576	929	100	657
10 Transport equipment	758	126	505	780	127	443	824	129	428
Construction	4	52	16	3	35	18	2	63	15
Trade & catering	0	426	0	1	504	0	2	613	1
Transport & comm.	63	190	68	63	159	75	75	202	106
Business services	244	236	162	149	157	327	227	251	597
Personal & public servs	4	123	1	6	207	0	10	473	1
Manufacturing	2310	842	1689	2637	781	1601	3173	828	1816
Services	312	976	231	219	1028	402	315	1539	704
Total	2759	2042	2042	3027	2128	2128	3693	2690	2690

feeders for other sectors, the interesting feature is the rise of the business services sector, which by 1990 had become a major 'supplier' of R&D although still only a small R&D spender. The rapid growth in demand for business services has meant that, in terms of the total amount of R&D embodied in its output, it had jumped to second place. By 1990 this sector was transmitting £597 million worth of R&D spending, compared with £657 million for electrical equipment, £428 million by transport equipment and £408 million for chemicals. As with the analysis of labour productivity this points to the need to differentiate high vs. low technology sectors within both manufacturing and services rather than equating stationary technology with services and dynamic technology with manufacturing.

Similar results on the role of services sectors in transmitting R&D through their purchases of intermediate and capital goods have recently been reported by the OECD for ten leading industrialised countries in the early 1990s (OECD, 1996; also Sakurai, Papaconstantinou and Ioannidis, 1997). Our estimates reinforce the results on the role of capital goods purchases in the transmission of R&D outputs between manufacturing and services found by Wolff and Nadiri (1993) for the US up to the 1970s; our period has the advantage of including the era of what has been termed the paradigm shift from electro-mechanical to microchip-based office technology. Our findings extend earlier work by Geroski (1991) which established the importance of the inter-industry spread of innovation within UK manufacturing and our results for user industries broadly match the findings by Scherer for the US, again in the 1970s (Scherer, 1984). However, we find lower rates of spreading from the producer industries as a proportion of their R&D than in Scherer's work, possibly due to the larger 'leakage' of R&D through exports from the UK. This greater openness to trade, while reducing the feed-through of the UK's domestic R&D, also enhances the product quality of inputs by improvements made elsewhere which are not captured by our analysis (on this see OECD 1996).

Conclusions

By using the complete accounting framework provided by the input–output statistics together with R&D data matched by sector, we have been able to trace the sources and impacts of various aspects of the process of technological change in the Thatcher era. The search for key sectors, generating cost savings and product improvements which spread throughout the economy, has led to a re-evaluation of the relative contributions of manufacturing and services to productivity growth. A general feature of this analysis is the demonstration of the symbiosis between manufacturers and service providers in the advanced industrial economy; developments in value added in any one sector are clearly not independent of the evolution of the quantity and quality of supply in others.

In the analysis of labour productivity growth we found the dynamic areas of the services sector, notably business services, to be key elements in the productivity record of the 1980s. This was due to their own high rate of productivity growth, combined with the very rapid expansion in their role as suppliers to the rest of the economy. More widely, we found dynamic and sluggish elements on both sides of the 'goods vs services' divide, confirming that key sectors have to be identified at a more disaggregated level than these conventional distinctions.

From the R&D analysis, tracking contributions to rising product quality, we revealed that high technology manufacturing and services bear more similarities to each other than high and low technology manufacturing. While the presence of the big R&D spenders in manufacturing ensures that product quality improvements are mostly generated there, the intersectoral transfer of these across sectors through sales of improved inputs facilitated efficiency gains and improvements in product and service quality much more widely. In particular the business services sector has become an important player in the forward transmission of rising product quality. If more accurate indicators of innovation in services were available, for example measuring copyright in information technology and software, these would enhance the estimated contribution of high-technology services, whose contribution to improved product quality is likely to be underestimated by the patterns of declared R&D.

Data Appendix: sources and adjustments

Input–output tables

Input–output tables for the UK are available for 1979, 1984, 1985 and 1990 (BSO, 1983; CSO, 1988, 1989, 1994). All are at current prices only, and are based on the 1980 Standard Industrial Classification. We have used the tables for 1979, 1985 and 1990. Although the 1985 tables are essentially an update of the benchmark tables for 1984, the miners' strike of 1984–5 introduced distortions into purchases of materials and fuels over that period. New estimates adjusting for this were made for 1985, that year having the further advantage of being a base year for constant-price output series, and cyclically more comparable with 1979 and 1990.

For 1979 both industry by industry and commodity by commodity domestic use tables are available, but for the later years only the commodity by commodity basis is available, and this has been used throughout. It is also the preferable basis, corresponding more closely to the homogeneous product and common technology assumptions of input–output analysis. Although the tables for all three years are based on the same Standard

Industrial Classification they contain minor differences in commodity aggregation: 100 sectors in 1979, 102 in 1985 and 123 in 1990. A maximum of 97 sectors could be achieved through direct aggregation. This was further reduced to 87 by the limited availability of sectoral deflators, particularly within the engineering sectors.

The GDFCF-published matrices contain purchases of domestically produced and imported capital goods combined, with the domestic/import split available only for each sector's total purchases. The proportions from the row sum was applied to each element in the sector (row) to obtain an estimate of the domestically produced capital matrix. Since the GDFCF matrix is supplied only on an industry by commodity basis, these also had to be converted to a commodity by commodity basis by application of the 'make' matrix.

Conversion to constant prices

The 87-sector current price data from CSO were deflated to a common 1985 price basis by sector-specific deflators. For the 76 primary and manufacturing sectors, including construction, we were able to use producer price indices and import average value indices which had been compiled by Oxford Economic Forecasting with assistance from CSO. To derive deflators for the 11 categories of domestically produced and imported services a more piecemeal approach had to be adopted. For domestic output of the three categories of financial services, other (mainly private sector) services and public services we derived implicit deflators from the CSO current and constant price net output data given in the Blue Book. (For the public sector this required the weighted aggregation of deflators from subheadings within this sector.) Although formally these implicit deflators relate to net output, the CSO indicate that many of the indicators used in the construction of the constant price series for net output are in practice gross output measures (CSO, 1985).

Similar implicit deflators are available for distribution, hotels and catering, which we distinguish as two sectors, and for transport and communications, where we distinguish six separate sectors. Our overall approach was to refine these to the more disaggregated level which we required. We first adopted the two more aggregated deflators as interim estimates for each of the two and six constituent sectors respectively. For the eight disaggregated sectors this completed the 87-sector vector of their inter-industry purchases at constant prices. The addition of sectoral employment income and gross profits, deflated as described below, gave an interim estimate of constant-price gross output for each of these sectors, derived through the constant-price gross output for each sector, derived as the column-total of its purchases. Juxtaposed with the current-price valuation of gross output this measure provided our final estimate of the implicit deflator for domestic output in each of the eight sectors.

Deflation of employment income

Employment income generated in each sector was converted to a 'constant 1985 price' basis by revaluing at 1985 sectoral hourly earnings. The method developed is analogous to the deflation of gross output. The quantity units of employment, or weights, in each sector were the total person-hours worked there in 1985 by up to six groups: full-time manual and non-manual workers, male and female, and part-time female manual and non-manual workers. Sector-specific data on hourly earnings for each of the six groups from the New Earnings Survey provided the 'price relative'. Combining these gave the index of earnings change for each sector, which was then used to deflate employment income. Our measure of real employment income thus represents the employment income generated in each sector, revalued at 1985 earnings per person-hour in that sector.

The fifteen-sector aggregation

After all matrices had been constructed and deflated at the highest level of disaggregation possible we proceeded with aggregation to 15 sectors for computation of results; the aim of this aggregation was to rebalance the data between the high degree of detail available for manufacturing, which is about one quarter of the economy, and to achieve a level suitable for matching with R&D data. Short descriptions of the 15 sectors used are given below, along with their composition in terms of both the SIC80 and SIC92 classifications.

The R&D data

The data used was UK business expenditure on R&D (BERD), obtained from the CSO Business Monitor series MO14 entitled 'Research and Development in UK Business'. R&D data are provided on a nominal basis by 33 sectors organised according to the SIC92. They were deflated according to the above method, that is, each sector was deflated using an individual sectoral R&D deflator calculated by Cameron (1996). The concordance used to map the data on SIC92 into the 15 sectors on SIC80 follows.

SIC80–SIC92 Mapping

Sector Full description
1 Agriculture, hunting, forestry & fishing
2 Extractive industries, utilities, oil & nuclear processing
3 Manufacture of food products, beverages & tobacco
4 Manufacture of pulp, paper & paper products, publishing & printing
5 Manufacture of basic metals & fabricated metal products
6 Manufacture not elsewhere classified
7 Manufacture of machinery & equipment not elsewhere classified
8 Manufacture of chemicals, chemical products & man-made fibres
9 Manufacture of electrical & optical equipment
10 Manufacture of transport equipment
11 Construction
12 Wholesale and retail trade, repair, hotel & restaurants
13 Transport, storage & communication
14 Financial intermediation, real estate, planning & business activities
15 Public administration & defence, social services, education, other services

Sector	Description used in figures	SIC80	SIC92
1	Agriculture	01,02,03	01,02,05
2	Energy & utilities	11,12,13,14,15, 16,17,21,23	10,11,12,13,14, 23,40,41
3	Food & drink	41,42	15,16
4	Paper & printing	47	21,22
5	Metals	22,31	27,28
6	Manufacturing nes	24,43,44,45,46, 48,49	17,18,19,20 25,26,36,37
7	Mechanical engineering	32	29
8	Chemicals	25,26	24
9	Electrical equipment	33,34,37	30,31,32,33
10	Transport equipment	35,36	34,35
11	Construction	50	45
12	Trade & catering	61,62,63,64,65, 66,67	50,51,52,55
13	Transport & communications	71,72,74,75,76, 77,79	60,61,62,63,64

| 14 | Business services | 81,82,83,84,85,94 | 65,66,67,70,71, 72,73,74 |
| 15 | Personal & public services | 91,93,95,92,96, 97,98,99,00 | 75,80,85,90 |

Note

1 We are very grateful to Ben Zissimos for research and computing assistance throughout the preparation of this paper. Ian Sue Wing also provided valuable computing assistance in the early stages of the work. Gavin Cameron has generously allowed us to use his R&D data, including price deflators, and Oxford Economic Forecasting made available their industry price deflators. The work is financially supported by the Leverhulme Trust and forms part of the Oxford University Institute of Economics and Statistics project on 'The Labour Market Consequences of Technical and Structural Change'. Useful comments were received from participants at the European Economic Association Annual Conference, Toulouse, Septemebr 1997, and at the NIESR/ESRC Conference on Productivity and Competitiveness, London, February 1998, particularly from Nick Oulton and Stan Metcalfe.

References

Bean, C. and Crafts, N. (1996), 'British economic growth since 1945: relative economic decline ... and renaissance?', in Crafts, N. and Toniolo, G. (eds), *Economic Growth in Europe since 1945*, Cambridge, Cambridge University Press.

Bernard, A.B. and Jones, C.I. (1996), 'Comparing apples to oranges: productivity convergence and measurement across industries and countries', *American Economic Review*, 86, 5, pp. 1216–38.

Blair, P.D. and Wyckoff A.W. (1989), 'The changing structure of the US economy: an input–output analysis', in Miller, R.E., Polenske, K.R. and Rose, A.Z. (eds), *Frontiers of Input–Output Analysis*, New York and Oxford, Oxford University Press.

Business Statistics Office (BSO) Department of Industry (1983), *Input–Output Tables for the United Kingdom 1979*, Business Monitor PA 1004, London, HMSO.

Cameron, G. (1996), 'Divisia price indices for UK business enterprise R&D', *Science and Public Policy*, October.

Carter, A.P. (1970), *Structural Change in the American Economy*, Cambridge, Mass., Harvard University Press.

Central Statistical Office (CSO) (1988), *Input–Output Tables for the United Kingdom 1984*, London, HMSO.

 (1989), *Input–Output Tables for the United Kingdom 1985*, London, HMSO.

(1994), *Input–Output Tables for the United Kingdom 1990*, London, HMSO.
(1996), *Research and Development in UK Business: Survey of Business Enterprise R&D 1993*, cf BSO Business Monitor MO14, London, HMSO.
Geroski, P. (1991), 'Innovation and the sectoral sources of productivity growth', *Economic Journal*, 101, p. 409.
Gregory, M. and Greenhalgh, C. (1997), 'International trade, de-industrialisation and labour demand – an input–output study for the UK 1979–90', in Borkakoti, J. and Milner, C. (eds), *International Trade and Labour Markets*, Basingstoke, Macmillan, chapter 4.
Griliches, Z. (1979), 'Issues in assessing the contribution of research and development to productivity growth', *Bell Journal of Economics*, 10, 1.
(1990), 'Hedonic price indices and the measurement of capital and productivity', in Berndt, E.R. and Triplett, J.E. (eds), *Fifty Years of Economic Measurement*, National Bureau of Economic Research Studies in Income and Wealth, 54, Chicago, University of Chicago Press.
(1995), 'R&D and productivity: economic results and measurement issues', in Stoneman, P. (ed.), *Handbook of the Economics of Technological Change*, Oxford, Basil Blackwell.
Griliches, Z. and Lichtenberg, F. (1984), 'Interindustry technology flows and productivity growth: a re-examination', *Review of Economics and Statistics*, 66, 2.
Grossman, G. and Helpman, E. (1996), *Innovation and Growth in the Global Economy*, Cambridge, Mass., MIT Press.
Hall, B. (1993), 'R&D tax policy in the eighties: success or failure?' NBER Working Paper 4240; published in *The Tax Policy and the Economy*, Cambridge Mass., NBER.
Hall, B. and Mairesse, J. (1995), 'Exploring the relationship between R&D and productivity in French manufacturing firms', *Journal of Econometrics*, 65.
Hart, P.E. (1996), 'Accounting for the economic growth of firms in UK manufacturing since 1973', *Cambridge Journal of Economics*, 20, 2.
Hulten, C.R. (1978), 'Growth accounting with intermediate inputs', *Review of Economic Studies*, 45, pp. 511–18.
Jorgenson, D.W., Gollop F.M. and Fraumeni, B.M. (1987), *Productivity and US Economic Growth*, Cambridge, Mass., Harvard University Press.
Jorgenson, D.W. and Griliches, Z. (1967), 'The explanation of productivity change', *Review of Economic Studies*, 34, 3.
Leontief, W.W. (1951), *The Structure of the American Economy 1919–39: An Empirical Application of Equilibrium Analysis*, New York, Oxford University Press.
Muellbauer, J. (1991), 'Productivity and competitiveness', *Oxford Review of Economic Policy*, 7, 3.
OECD (1996), *Technology and Industrial Performance: Technology Diffusion, Productivity, Employment and Skills, International Competitiveness*, Paris, OECD.
Oulton, N. and O'Mahony, M. (1994), *Productivity and Growth: A Study of British Industry, 1954–86*, Cambridge, Cambridge University Press.
Postner, H. and Wesa, L. (1983), 'Canadian productivity growth: an alternative (in-

put–output) analysis', Ottawa, Economic Council of Canada.
Rowthorn, R. and Wells, J. (1987), *Deindustrialisation and Foreign Trade*, Cambridge, Cambridge University Press.
Sakurai, N., Papaconstantinou, G. and Ioannidis, E. (1997), 'The impact of R&D and technology diffusion on productivity growth: empirical evidence for ten OECD countries', *Economic Systems Research*, 9, 1.
Scherer, F.M. (1982), 'Inter-industry technology flows and productivity growth', *Review of Economics and Statistics*, 54, 4.
 (1984), *Innovation and Growth: Schumpeterian Perspectives*, Cambridge, Mass., MIT Press.
Solow, R. (1957), 'Technical change and the aggregate production function', *Review of Economics and Statistics*, 70, pp. 312–20.
Terleckyj, N.E. (1974), *The Effects of R&D on the Productivity Growth of Industries: An Exploratory Study*, National Planning Association, Washington DC.
 (1980), 'Direct and indirect effects of industrial research and development on the productivity growth of industries', in Kendrick, J.W. and Vaccara, B.N. (eds), *New Developments in Productivity Measurement and Analysis*, National Bureau of Economic Research Studies in Income and Wealth, 44, Chicago, University of Chicago Press.
Wolff, E.N. (1996), 'The productivity slowdown: the culprit at last?', *American Economic Review*, 86, 5, pp. 1239–52.
Wolff, E.N. and Nadiri, M.I. (1993), 'Spillover effects, linkage structure and research and development', *Structural Change and Economic Dynamics*, 4, 2.

4 Productivity, employment and the 'IT paradox': evidence from financial services

MICHELLE HAYNES AND STEVE THOMPSON[1]

1 Introduction

There is considerable current academic interest in productivity growth and employment in the financial services industry. This is not simply a reflection of the secular increase in importance of the industry itself, but also recognises that financial services have been experiencing, albeit at an accelerated rate, fundamental changes in technology and in the liberalisation of competition which in time may come to impact upon much of the service sector. Given the simple proportionate size of the latter in modern, post-industrial economies, it is clear that service productivity and employment constitute important structural issues. This chapter seeks to address some of these by an examination of the impact of a major embodied IT innovation as it diffused through a set of financial services firms.

The chapter is partly motivated by what has become known, particularly in the US literature, as the 'IT productivity paradox' (Berndt and Malone, 1995, Brynjolfsson, 1993, Brynjolfsson and Hitt, 1996, Wilson, 1995 and so on). This term relates to the apparent failure of so much empirical work to demonstrate a clear productivity impact resulting from very high levels of IT investment. Or, in the words of Robert Solow quoted by Catherine Morrison (1997, p. 471): 'We see computers everywhere except in the productivity statistics.'

It is clear that many of the studies which have attempted to assess the impact of IT inputs on firm or industry outputs have encountered severe measurement problems. IT inputs are frequently difficult to separate from other investment expenditures and often manifest considerable cross-sectional and time-series heterogeneities. Output measurement is substantially complicated, as Zvi Griliches has shown in his work on the computer industry, by the input-induced changes in the hedonic characteristics of the output. Here we seek to circumvent some of these problems by examining the impact of a single homogeneous embodied IT innovation, the automated teller machine (ATM), across a sample of potential users operating in the same core business. It is suggested that the ATM is typical of many embodied IT applications in that it combines elements of process and product innovations. We then examine the impact of ATM adoption on labour productivity and then labour demand, using an unbalanced panel of 93 UK building societies over the years 1981-93, a period corresponding to that of rapid diffusion of new technology in the sector.

Our empirical results suggest that ATM adoption had a large and statistically significant impact on both productivity and the demand for labour. While capital measurement anxieties precluded total factor productivity analysis, our examination of labour productivity suggested that the introduction of an ATM system produced an immediate and sustained increase in per capita output. By the early 1990s this amounted to a mean difference of approximately 26 per cent. The paper then uses a dynamic labour demand model, augmented with variables to capture the contemporaneous and lagged employment effects of ATM introduction. This model is estimated on an unbalanced panel, using a generalised method of moments procedure to allow the inclusion of a lagged dependent variable. The labour demand model is well determined, with elasticities of appropriate signs and magnitude. While different specifications were employed to accommodate ATM introduction, the results indicated large, significant, contemporaneous and lagged negative effects on adopters' labour demand. When the adopters were further dichotomised into intensive and non-intensive users of the technology it was apparent, as expected, that the former displayed a greater *ceteris paribus* fall in labour demand than the latter; although even the non-intensive users showed significant negative effects.

Thus our results on ATM introduction into financial services give no support to the IT productivity paradox. Indeed, they suggest that adopters of embodied IT innovations may enjoy very substantial labour savings by comparison with their non-adopting rivals, at least over the period of early diffusion. However, it is important to note at least two important caveats to our results: First, as with any innovation, those users with the greatest potential benefit from introduction will be disproportionately represented among the early adopters, implying that their gains probably overestimate the consequences of full diffusion across the sample; and second, as Geroski, Machin and Van Reenan (1993) have pointed out in a similar context, it is possible that at least some of the apparent firm-specific consequences of innovative activity are actually attributable to superior (that is, more innovative) management.

The remainder of the chapter is structured as follows: section 2 discusses information technology investment in the financial services industry in the context of the wider literature on the IT productivity paradox and considers ATM adoption as an example of such investment. Section 3 describes the data and provides a preliminary examination of the productivity effects of ATM adoption. The dynamic demand for labour model is outlined in section 4, and section 5 presents the empirical results. A brief conclusion follows.

2 Financial services, the IT productivity paradox and the ATM

2.1 Financial services and IT investment

The apparently elusive productivity contribution of information technology investment to the economy, the substance of the so-called 'IT paradox', remains a highly controversial issue. However, as empirical evidence accumulates and measurement issues are resolved, the judgement on manufacturing industries is starting to appear more sanguine (see below) and the debate is shifting to the service sector. This is hardly surprising. Service industries in general, and financial services in particular, have been disproportionately large investors in IT over the past fifteen or so years. This position appears even more pronounced when viewed against the traditionally low levels of fixed asset investment in services. Thus Roach (1991), for example, estimates that by 1989 the service sector accounted for 85 per cent of the accumulated US information technology investment, while IT constituted 45 per cent of the US banking industry's total capital stock. Griliches (1994) notes that by 1992 computer related investment in finance, insurance and real estate amounted to 37.8 per cent of the annual total for the US economy. Similarly, a sectoral comparison by Brynjolfsson and Hitt (1996) finds that the ratio of IT expenditure to the non-IT capital stock for US financial services firms exceeds that for the all-sector average by a factor of six.

Critics have argued that service sector productivity has been particularly unresponsive to these IT expenditures and, indeed, that many service sector firms have been merely profligate with shareholders' resources. Stephen Roach, Senior Economist at Morgan Stanley, provides a trenchant statement of the this view:

> The service sector's lagging efficiency may seem surprising in the light of the costly bet that has been wagered on the promise of computers, telecommunications and other forms of information technology to enhance productivity. While economists have long recognised the relationship between capital endowments and productivity enhancement, this relationship has not been borne out in services. The massive investments in technology simply have not improved productivity, on the contrary they have made service organisations less profitable and less able to compete . . . (Roach, 1991, p.85)

While Roach does concede a few exceptions to his generalisations on the unproductive nature of IT investment in services, the financial services industry is certainly not one of them. On the banking industry he concluded:

> Investment in such (information) technology rose 20 per cent a year during the 1980s; by 1989, . . . labour costs had changed little. Employment growth slowed only slightly from 1 per cent a year to 0.3 per cent. In general, technology has not made banking operations more efficient and the cost structure remains bloated. (p. 90)

Comments such as these raise two issues: first, whether there is something intrinsically different about service sector production that results in an inevitably lower marginal product for IT investments; and second, whether it is the measurement difficulties attached to service sector research which conceal the true extent of any productive contribution from IT.

The first issue is an empirical one which is only just starting to be addressed. Brynjolfsson and Hitt (1996), for example, estimate production functions for samples of manufacturing and service sector firms using IT expenditures as a separable input. They conclude that whilst there is no general significant difference in the elasticities across the two samples, firm-specific differences are important. In particular, they suggest that, following David (1990), gains are differentially realised by those service sector firms which follow complementary strategies of restructuring.

The second issue concerns the more general problem of making output comparisons across sectors. It has been argued by Griliches (1994) that the service/manufacturing comparison is less important than that between measurable and unmeasurable outputs. It is, he suggests, precisely because service output largely falls into the latter category that comparisons of productivity growth are so deficient and it is because so much computer investment is also concentrated within the unmeasurable sectors that the apparent gains from IT deployment are so small.

2.2 The IT productivity paradox debate

In one sense the debate over service sector productivity is merely the latest twist in a longer-running controversy over the productivity of computer investment. The term IT productivity paradox has been coined to describe the incongruency between the obvious transformation which the information technology revolution is bringing in work practices and the failure of this change to be reflected more clearly in the output statistics. If today's IT revolution is as fundamental as were, say, the eighteenth century agricultural and industrial revolutions, as some have suggested (for example, Malone and Rockart,1991, David,1990 and so on), it might be expected that this would be reflected more clearly in the evidence. This has been slow to happen.

Taken as a whole, the empirical evidence on the output effects of IT investments at the macro economy, industry and firm levels of aggregation remains ambiguous and is almost exclusively restricted to work in the United States. At the macro level it has been reasoned that economies such as the US, which enjoyed an obvious lead in the adoption of IT, should demonstrate differential productivity gains in the period of its diffusion. In fact, measured productivity growth is disappointingly stable and attempts to correlate it with IT investment (see Roach, 1992) have proved unsuccessful. Of course, the output of the computer industry is itself a major addition to GDP in many countries and one which becomes enormously more important when a hedonic price deflator is used.[2] Even here, however, as Jorgenson and Stiroh (1995) point out, there is a severe measurement problem: the dramatic price falls for IT hardware simultaneously create difficulties on the input side where the capital values of acquired equipment fall sharply and where the real flows of IT input services rise rapidly, if a hedonic deflator is used.

A recent literature survey by Wilson (1995) covers twenty industry- and firm-level studies of the impact of IT investments. While these vary substantially in coverage and econometric sophistication, it is notable that seven studies generally recorded a positive association between IT inputs and measured outputs, while thirteen found disconfirming evidence. Among the industry level studies, Morrison and Berndt (1990) (see also Morrison, 1997), using a production function approach, actually report a significant negative net marginal benefit from computer investment. A subsequent study by Berndt and Morrison (1994), which uses the proportion of 'high tech' office equipment to total capital as a measure of IT input, finds a positive association between IT inputs and profitability but a negative relationship between IT and both labour and multifactor productivity.

Among the firm level studies, which cover a variety of US industries, Loveman (1994) reports insignificant marginal returns to IT across a sample of sixty firms over the period 1977–82, the early years of the technology's diffusion. Barua, Kriebel and Mukhopadhyay (1991), also using US firm-level data, find no correlation between computer investment and financial performance.

Possible explanations for the IT paradox may be grouped under four main headings: inadequate measurement of variety and quality gains, extended lags in full implementation, redistribution to consumers and mismanagement. These may be considered briefly in turn:

a) *Measurement of quality and variety.* It is clear that many – and perhaps most – IT applications are neither pure process nor pure product innovations. Process IT innovations typically allow entirely new or substantially improved characteristics to be incorporated in the delivery of existing products – for example in telecommunications – and/or allow enhanced product variety, as in the application of flexible manufacturing systems. It is arguable that existing output accounting systematically understates the gains from such developments, as compared, say, with one using hedonic price indices.

b) *Extended lags.* It is arguable that to exploit the benefits of IT to the full requires the development of co-specialised assets, most obviously in human capital, and the reorganisation of firms and, perhaps, contracting patterns. Such developments may involve longer lags than the adoption of the technologies themselves. David (1990) has developed this point using the example of electrification in the 1930s which, he suggests, did not yield its full benefits until factories were redesigned to accommodate the more flexible electrical equipment.

c) *Redistribution to consumers.* Product competition between IT users in any particular market will be expected to drive down prices leading to a redistribution of gains towards consumers. This will be particularly pronounced if the IT application lowers entry barriers to the market.

d) *Mismanagement.* Finally, some researchers have argued that the potential benefits of IT adoption have been squandered because of poor design and deployment or inappropriate investment (Wilson, 1995). Proponents of the mismanagement view often supply anecdotal support, but systematic evidence appears more problematic.

However, it is notable that some of the most recent US work on manufacturing presents a less pessimistic picture. Brynjolfsson and Hitt (1996) use data from 367 large US firms to estimate a production function equation for five years, 1987–92, using a seemingly unrelated regression design. They disaggregate capital and labour inputs into IT and non-IT components and report very substantial returns for the former, a result which appeared robust to different specifications of the production function. Lichtenberg (1995), using a similar data set, provides confirmation of substantial firm-level returns to computer investment.

While there is a substantial general literature on the impact of new technology on employment (see the review by Meyer-Krahmer, 1995), very few of the contributors to the debate on the IT productivity paradox have attempted to estimate the direct employment effects of IT adoption. Osterman (1986), in a very early study using production function estimates for forty US service sector firms, reported small but significant displacement effects: for example, his elasticities of employment with respect to computer application were –0.18 and –0.12, for clerical and managerial employees respectively.[3] Brynjolfsson *et al.* (1994), adopting a more indirect approach, have recently examined the impact of (sectoral) IT inputs on firm size, including employment, using a sample of 363 large US corporations. The authors find that large IT investment is associated with a significant decline in employment per establishment, following a one- to two-year lag. However, the authors find a very similar set of coefficients for other size measures including sales per establishment which, as they indicate, suggests that organisational and perhaps contractual changes are involved as well.

It appears likely that many of the apparently contradictory conclusions in the studies reviewed here result from problems in measuring IT inputs. Firm- and industry-level studies rarely have access to investment expenditure disaggregated to the type of capital inputs purchased. Even where it is possible to use reasonably precise categories of expenditure – for example, office equipment – it is likely that there are substantial inter-industry and inter-firm differences in the technological composition of the equipment purchased.[4] As previously noted, there are particular difficulties attached to the quantification of IT inputs through time, when price falls and quality changes create the need for a hedonic deflator but simultaneously generate valuation problems for past investments.

Thus the measurement of IT capital inputs is plagued with problems of cross-sectional and time-series heterogeneities. The remainder of this chapter seeks to circumvent some of these problems by concentrating upon the observable impact of the deployment of the ATM, which is treated as a single homogeneous embodied application of IT.

2.3 The ATM as an IT application

The ATM is typical of many embodied applications of IT in that it possesses attributes of both process and product innovations. It is clearly a process innovation, which may be defined as involving 'a change in the way products are made and delivered' (Tushman and Nadler, 1986), since by definition it substitutes the automated delivery of services for their provision by a bank teller. Furthermore, research on the diffusion of ATMs in both the US and UK confirms the positive impact of the wage rate on the speed of adoption, as would be expected for a labour saving innovation (Hannan and McDowell, 1984; Ingham and Thompson, 1993). However, in addition it offers improved services or even characteristics not previously offered by financial intermediaries: for example, 24-hour access, foreign currency provision abroad and so on. As such it constitutes a product innovation, with implications for consumer demand. For example, Daniels and Murphy (1994) show that access to an ATM network reduces customer currency holdings and correspondingly increases transactions balances while, as Griliches (1994) points out, such access generates unmeasured but not unvalued time savings for the users.

In common with many other IT innovations the ATM exhibits important network externalities (Salop, 1990). The more locations available to the user the more valuable is access to the system. In consequence, independent financial intermediaries have co-ordinated their introduction of ATMs into a series of joint networks, many of which have themselves developed access linkages. Therefore once an institution has introduced the technology and joined the network its customers will have access to all the eligible ATM sites so that the bank will not necessarily find it advantageous to provide a comprehensive coverage of ATMs at its own branches. As a result it might be anticipated that banks will differ not simply with respect to the decision about when to adopt the new technology but also with respect to the desired intensity of adoption.

The ATM remains within a service industry characterised by the use of the product at the point of delivery. This has the important implication that its utilisation depends directly upon consumer demand.[5] By contrast, the utilisation of many IT applications in manufacturing will be initially determined by production scheduling of the operator. It follows that during the diffusion stage of the ATM, when potential users may be unaware of its attributes or unable to access it, the installed transactions capacity might be expected to exceed the level of utilisation.

3 Data and preliminary analysis

3.1 Data and variables

The sample consisted of an unbalanced panel of ninety-three UK building societies over the years 1981–93, inclusive. The period was chosen to correspond with that of rapid diffusion of the ATM across the sector, see table 4.1.[6] To qualify for inclusion it was necessary that: (a) at least five years of continuous data were available; and (b) that at least two years of observations related to the post-deregulation regime which commenced in 1987. These requirements excluded some (generally small) societies which were acquired over the period. Since the population as a whole fell from approximately two hundred at the start of the period to eighty at the end, it can be seen that our sample was very close to the extant population in the latter part of the study. The temporal characteristics of the unbalanced panel are shown in table 4.2.

Data on the adoption of ATMs were obtained from the annual surveys of new technology published by *Banking World* and from the *Building Societies Association Yearbook* (various issues). Other data were obtained from industry sources, including the Building Societies Association, and annual reports and accounts held at Franey and Co., London, Manchester Central Reference Library, Loughborough University Banking Centre Library and Warwick University Library. As far as possible merger activity was allowed for with a continuous treatment of the data for the identified acquiring society. In only one case did complex merger activity prevent the inclusion of an adopting society. A particular advantage to the data concerns the regulation and accounting treatment of diversification. In other firm-level studies of IT adoption the adopters will typically include a substantial proportion of multi-output firms, such that the observed consequences of adoption will be a conflation of separate effects. UK building societies, by contrast, were limited to core activities of deposit collection and mortgage finance until 1987 and thereafter were permitted strictly limited diversification with the requirement that separate accounts were maintained for the core and group activities. Therefore by restricting attention to the societies' core accounts it seems reasonable to assume a homogeneous activity across the sample.

The intra-sample diffusion of the ATM as an innovation is illustrated in table 4.1. A total of twenty-three out of ninety-three sample firms adopted ATMs in the period of investigation. The adopters varied considerably in the extent of their deployment of ATMs – from a maximum of 1,600 units to a minimum of two – and in the intensity of ATM usage

Table 4.1 *Diffusion of ATM technology across the sample*

Year	Percentage of adopters
1981	0
1982	0
1983	1
1984	2
1985	5
1986	14
1987	19
1988	22
1989	26
1990	26
1991	27
1992	28
1993	27

Table 4.2 *Balance of panel*

Number of years	Number of companies
5	2
6	1
7	2
8	4
9	1
10	8
11	14
12	23
13	38
Total	93

relative to, say, the number of branches.[7] However, the intra-society diffusion pattern showed a considerable similarity across the sample with most adopters getting close to their maximum number of installed ATMs within a few years of the initial adoption: for example, eleven of the adopters had reached 90 per cent of their maximum within four years and another eight had reached 50 per cent in the same time.

Table 4.3 *Full period characteristics for continuous variables*

Variable	Sample (n = 1075)		Adopters (n = 169)		Non-adopters (n = 906)	
Wage	0.01	0.00	0.01	0.00	0.01	0.00
Employment	706.22	1919.80	2990.16	3688.71	280.18	831.99
Output (£m)	1236.59	3629.02	5656.99	7193.17	412.40	1303.24

Notes: Number of firms = 93; number of adopters = 23; of which: intensive users = 9, non-intensive users = 14.

The definition of the variables used in the analysis is as follows:

Total employment (N_{it}). Total employment is measured by the number of full-time equivalent employees of society i at time t, where part-time employees are treated as half full-timers.

Wages (W_{it}). The average wage per employee of society i at time t is calculated as total labour costs, expressed in 1985 prices, divided by the number of full-time equivalent employees.

Output (Q_{it}). We have followed other researchers in taking the intermediation view of bank production, see Sealey and Lindley (1977), and hence treating the firm's output as its level of earning assets. Accordingly, output is calculated by taking the balance sheet value of commercial assets, in constant 1985 prices, at time t.

ATM adoption (ATM_{it}). This is measured as a dichotomous variable equal to 1 in the year of adoption and 0 otherwise. The subsequent effects of the intra-firm diffusion of the ATM are captured by lagged values of ATM_{it}.

Intensive and non-intensive ATM users. To distinguish intensive ATM users, which might be expected to display major labour saving effects, from non-intensive users, we used the ratio of installed ATMs to branches after four years of ATM operation. (As noted above, intra-firm diffusion typically proceeded at a rapid pace after commencement of ATM installation; whilst, at least for the period of investigation, ATMs were overwhelmingly located at the branches). Societies with an ATM/branch ratio greater than 0.33 were deemed intensive users (*ATMI*) (nine cases) and those adopters with a ratio less than that were deemed non-intensive users (*ATMNI*) (fourteen cases).

The summary statistics for the continuous variables are given in table 4.3.

3.1 Productivity effects: a preliminary view of the data

In conformity with the intermediation model of bank production (Sealey and Lindley, 1977) the banking firm's output is taken to be its real level of earning assets which is produced from capital and labour inputs. Data permitting, this would allow the assessment of total factor productivity and hence make possible the calculation of the full productive impact of the ATM as an innovation. However, in this case there are limitations in the aggregate balance sheet capital data which, it is considered, make it less than satisfactory for this purpose. Accordingly, we initially present output per capita data comparisons for all adopters and non-adopters and then a comparison of the *changes* in output per head for adopters and non-adopters around the time of adoption.

Table 4.4 presents a comparison of yearly real per capita output (Q_{it}/N_{it}) for all the adopters and non-adopters in the sample. It can be observed immediately that once the number of adopters increased beyond two, in 1985, the adopting firms displayed a significantly higher output per head in every year of the comparison. By the early 1990s this differential was equivalent to over 25 per cent of the non-adopters' per capita output. Of course, as table 4.4 also illustrates, both the sample as a whole and the subsample of non-adopters in particular display high rates of attrition which may distort yearly comparisons. Therefore table 4.5 presents the mean change in per capita output for the adopters relative to the comparable period mean change for a constant sample of non-adopters. (These being all the societies which remained in existence at the end of the period but which had held out against adoption.) That is:

$$\Delta(Q_i/N_i)_{t-1,t+1} - \Delta(Qn/Nn)_{t-1,t+1} = [(Q_{it+1}/N_{it+1}) - (Q_{it-1}/N_{it-1})] \\ - [(Q_{nt+1}/N_{nt+1}) - (Q_{nt-1}/N_{nt-1})] \quad (1)$$

where Q_{nt}, N_{nt} equal mean values for a consistent set of non-adopters at time t, and so on.

By restricting attention to the *change* in adopters' output per head, table 4.5 avoids the scale bias which would arise as larger societies adopt first. Similarly, by restricting the comparison to the set of *surviving* non-adopters, the table avoids the effects of attrition bias in the control group. In the event, table 4.5 indicates that adoption was accompanied by a substantial and statistically significant increase in per capita output, even in comparison with the contemporaneous performance of those non-adopters which passed a survival test. Furthermore, the mean difference increases in the years following adoption and doubles between years t and $t + 2$.

It was recognised that the adopting firms could benefit from produc-

Table 4.4 *Real output per full-time equivalent employee*

Year	Non-adopters (n)		Adopters (n)		Difference	(t-ratio)
1981	1.15	(48)	n/a	(0)	n/a	–
1982	1.06	(73)	n/a	(0)	n/a	–
1983	1.13	(86)	1.31	(1)	0.18	–
1984	1.22	(88)	1.31	(2)	0.09	(0.82)
1985	1.22	(85)	1.63	(5)	0.41	(3.13)
1986	1.25	(79)	1.71	(13)	0.46	(5.22)
1987	1.32	(74)	1.66	(18)	0.34	(3.57)
1988	1.41	(71)	1.67	(20)	0.26	(2.43)
1989	1.38	(66)	1.68	(23)	0.30	(3.26)
1990	1.44	(64)	1.71	(23)	0.27	(2.63)
1991	1.44	(60)	1.81	(22)	0.37	(3.03)
1992	1.43	(58)	1.85	(22)	0.42	(3.76)
1993	1.48	(54)	1.93	(20)	0.45	(3.60)

Table 4.5 *Mean change in per capita output*

	Mean	t-value	n
$\Delta(Q_i/N_i)_{t-1,t} - \Delta(Qn/Nn)_{t-1,t}$	0.06	3.00	23
$\Delta(Q_i/N_i)_{t-1,t+1} - \Delta(Qn/Nn)_{t-1,t+1}$	0.08	2.67	23
$\Delta(Q_i/N_i)_{t-1,t+2} - \Delta(Qn/Nn)_{t-1,t+2}$	0.12	2.40	23

tivity-enhancing characteristics which, although correlated in some way with the adoption decision, were not themselves a consequence of it. For example, superior-quality managers could both instigate the adoption of an innovation and independently raise factor productivity or, equivalently, adoption could be part of a broader strategy of capital investment and/or production reorganisation. Some evidence on this was generated by a comparison of the productivity per head of non-adopters and adopters in the three years *prior* to the latter's adoption. In the event there were no statistically significant differences in years three and two prior to adoption, but relative labour productivity was higher in the year before adoption, although the difference was very much smaller than it would come to be in subsequent years.[8]

Thus our preliminary analysis of the data was consistent with the ATMs having a major impact on productivity, in apparent conflict with the IT paradox. The next stage of the analysis was to conduct a more rigorous investigation of the impact of adoption on the demand for labour across the sector.

4 Demand for labour: the model

We begin by deriving a standard labour demand equation from the Cobb–Douglas production function:

$$Q_{it} = A_{it} L_{it}^a K_{it}^b e^{ct} \tag{2}$$

where Q_{it}, L_{it} and K_{it} are, respectively, the output and labour and capital inputs of firm i at time t and e^{ct} is an exponential time trend.

We denote the wage rate w_i and rental cost of capital r_i and assume that observed employment (E) proxies for labour input (L) – that is, homogeneity of labour services is assumed. Then from the firm's first order conditions a demand equation for desired employment (E^*) is generated:

$$\ln E^*_{it} = \beta_0 + \beta_1 \ln Q_{it} + \beta_2 t + \beta_3 \ln(w_{it}/r_{it}) \tag{3}$$

Following the labour demand literature (see Hamermesh, 1993) it is assumed that the adjustment of actual employment to desired employment is mediated by costs of hiring and firing. Adopting a partial adjustment mechanism:

$$\ln E_{it} - \ln E_{it-1} = \lambda (\ln E^*_{it} - \ln E_{it-1}) \tag{4}$$

Thus:

$$\ln E_{it} = \beta_0 + \beta_1 \ln Q_{it} + \beta_2 t + \beta_3 \ln(w_{it}/r_{it}) + \beta_4 \ln E_{it-1} \tag{5}$$

In common with other studies in the banking production literature (see Hardwick and Ashton, 1997, for a survey) this research found the derivation of an adequate proxy for r_{it} to be problematic. Therefore we followed Murray and White (1980) and many other contributors who have followed a bank production function approach in assuming the rental cost of capital to be constant across the sample and accordingly normalised r to one.[9] Adding a stochastic error term to (5) and then first differencing to remove the firm-level fixed effects:

$$\Delta \ln E_{it} = \beta_1 \Delta \ln Q_{it} + \beta_2 \Delta t + \beta_3 \Delta \ln w_{it} + \beta_4 \Delta \ln E_{it-1} + U_{it} \tag{6}$$

The ATM is considered here as a fundamental and completely new innovation whose adoption causes an immediate impact on the production function for banking services. That is, the ATM is treated as a completely new, separable capital input. Since we are observing its impact over the period of rapid inter-firm diffusion – a period during which some sample firms may be experiencing disequilibrium in cost minimisation – it appeared appropriate to evaluate the consequences as a parametric shift. However, it was noted above that, unlike technological innovations in manufacturing industry, the utilisation of an ATM system depends directly upon customer behaviour, which itself is subject to some learning process, and upon the growth in network externalities. Therefore it is assumed that the full impact of introducing an ATM system on each society's demand for labour will occur with some lag as customer behaviour adjusts. However, we had no strong priors on the appropriate form of lag structure. Accordingly, alternative specifications of equation (6) were employed with different sets of dummy variables to capture the immediate and lagged effect of ATM adoption on labour demand.

It is well-established that first-differencing a dynamic model to eliminate individual effects introduces correlation between the error term and the lagged dependent variable (Nickell, 1981). Fortunately this problem can be tackled by instrumental variables techniques. An obvious candidate is to use lags beyond one of the dependent variables as instruments as these will be uncorrelated with the differenced error in the absence of serial correlation in the error process. We also utilise values of capital stock as additional instruments.

In order to estimate the parameters in equation (6), we utilise the generalised method of moments procedure, contained in Arellano and Bond's (1991) DPD programme. The advantage of this procedure is that it allows both the cross-section and time-series elements of the data to be exploited in constructing valid instruments, offering significant gains in efficiency. The resulting estimates will be consistent in the absence of serial correlation.

5 Demand for labour: results

The results of estimating equation (6), using Arellano and Bond's (1991) GMM estimator in a two-step procedure, are presented in table 4.6. The asymptotic t-statistics are given in parentheses and are robust to heteroscedasticity. As one cross-section is lost from first differencing and another three from the instrumentation process, our estimation period runs from 1985 to 1993, inclusive, covering a total of 703 useable observations. Since a number of societies were adopting ATMs at around the same date, time dummies were not included, although in some estimations a housing market recession dummy (see below) was introduced to capture the impact of the highly unusual conditions of the early 1990s property market recession.

Column (1) reports our estimates of the basic labour demand model augmented by contemporaneous and lagged ATM adoption variables. In column (2) the ATM adopters are dichotomised into intensive adopters (*ATMI*) and non-intensive adopters (*ATMNI*). Columns (3) and (4) replicate the initial estimations with the addition of the housing market recession variable. In columns (5) and (6) the set of ATM variables is replaced with a single binary variable, *ATMA*, equal to zero prior to adoption and one thereafter, for all societies.

A comparison of columns (1) and (2) shows that the estimated coefficients are similar in each equation. Both equations are well determined and have sensible properties for a labour demand model. That is the coefficients on lagged employment, wage and output fall well within the range of values typically reported in labour demand studies (see Hamermesh, 1993). In each case the overall regression diagnostics are satisfactory. A Wald test on the joint significance of all the regressors is overwhelmingly significant. Similarly, a second Wald test on the joint inclusion of the subset of ATM variables is comfortably significant at the 0.1 per cent level. The assumption of a lack of serial correlation in the error is essential for the consistency of the estimates. Table 4.6 therefore reports test statistics for the validity of the instrumental variables approach. A robust $N(0,1)$ test for the presence of second-order serial correlation in the error term (*p*-values recorded in table 4.6) is satisfactory. Similarly, a Sargan test of instrument validity does not reject exogeneity of the instrument set.

Turning to the principal variables of interest, in column (1), the ATM terms enter as predicted with a negative sign and are individually and jointly significant up to and including lag three.[10] While the impact elasticity (–0.052) was quite small, the long-run effect (= $Sg/1 - a$) was equivalent to a 57 per cent fall in employment for societies adopting an ATM system. Furthermore, it will be recalled that since equation (5) controls for output changes, the employment effects are measured *after* any output gains consequent upon the product innovation characteristics of the ATM.[11]

In column (2) the dichotomisation of adopters into intensive and non-intensive users confirms our conjecture that employment effects will be greater for the former. Indeed the long-run effect appears almost implausibly large, with intensive adopters showing a *ceteris paribus* 81 per cent fall in labour demand. However, it was the case that several of these societies had very large ATM/branch ratios (well over unity) and it is likely

Table 4.6 *Employment equations, 1985–93 (absolute asymptotic t-statistics in parentheses)*

	(1)	(2)	(3)	(4)	(5)	(6)
Constant	−0.008	−0.007	−0.003	−0.002	−0.002	0.005
	(1.653)	(1.288)	(0.650)	(0.503)	(0.388)	(1.166)
$\Delta \ln N_{i(t-1)}$	0.637	0.623	0.647	0.624	0.645	0.684
	(9.614)	(9.679)	(9.832)	(9.948)	(10.088)	(9.966)
$\Delta \ln W_{it}$	−0.452	−0.440	−0.441	−0.431	−0.538	−0.596
	(6.635)	(6.407)	(6.197)	(5.988)	(7.817)	(8.204)
$\Delta \ln Q_{it}$	0.515	0.504	0.478	0.479	0.521	0.490
	(10.662)	(9.772)	(10.643)	(10.602)	(10.652)	(10.399)
ATM_{it}	−0.052	−	−0.053	−	−	−
	(4.640)		(4.845)			
$ATM_{i(t-1)}$	−0.049	−	−0.051	−	−	−
	(3.514)		(3.696)			
$ATM_{i(t-2)}$	−0.055	−	−0.057	−	−	−
	(3.143)		(3.229)			
$ATM_{i(t-3)}$	−0.052	−	−0.055	−	−	−
	(2.680)		(2.817)			
$ATMI_{it}$	−	−0.048	−	−0.047	−	−
		(3.025)		(3.235)		
$ATMI_{i(t-1)}$	−	−0.076	−	−0.076	−	−
		(4.486)		(4.844)		
$ATMI_{i(t-2)}$	−	−0.092	−	−0.094	−	−
		(4.459)		(6.675)		
$ATMI_{i(t-3)}$	−	−0.093	−	−0.096	−	−
		(3.068)		(3.462)		
$ATMNI_{it}$	−	−0.050	−	−0.054	−	−
		(3.667)		(4.089)		
$ATMNI_{i(t-1)}$	−	−0.040	−	−0.044	−	−
		(2.192)		(2.528)		
$ATMNI_{i(t-2)}$	−	−0.039	−	−0.045	−	−
		(1.800)		(1.990)		
$ATMNI_{i(t-3)}$	−	−0.037	−	−0.041	−	−
		(1.497)		(1.705)		
ATMA	−	−	−	−	−0.111	−0.119
					(3.857)	(4.131)
90s RECESS	−	−	−0.016	−0.015	−	−0.020
			(1.542)	(1.474)		(1.166)
Wald(all)	2267.6	2598.9	2410.2	2692.9	2399.9	2507.1
[df]	[8]	[12]	[9]	[13]	[5]	[6]
Wald(all ATM)[df]	24.01 [4]	54.02 [8]	25.62 [4]	55.51 [8]		
Serial correlation [df]	−0.641 [90]	−0.597 [90]	−0.674 [90]	−0.623 [90]	−0.521 [90]	−0.405 [90]
Sargan [df]	42.39 [32]	41.85 [32]	42.20 [32]	42.03 [32]	40.29 [32]	38.95 [32]
No. of firms	93	93	93	93	93	93
No. of obs.	703	703	703	703	703	703

that ATM adoption is correlated with other unobservable IT inputs, perhaps ones drawing upon the same computer systems. It is also the case that since the ATM exhibited product innovation characteristics as well as those of a process innovation – typifying IT applications – it will have a direct impact on our output variable through strategic competition within the industry. This may have produced a particularly strong growth in adopters' output during a period of rapid industry growth in the 1980s.

The non-intensive adopter variables have negative coefficients which are significant up to the second lag. However, it can be seen that the lagged effect of adoption is very much smaller than that for the intensive user. The long-run effect of adoption here is equivalent to a *ceteris paribus* 44 per cent fall in employment.

Replacing the set of ATM variables with a single binary variable (*ATMA*) for all adopters again produced a substantial, significant, negative effect. While the magnitude of this coefficient may appear surprisingly large, it must be remembered that the distribution of adoptions means that there are relatively small numbers of post-adoption years for most societies.

Finally, it was recognised that our estimation interval included a period of chronic recession in the housing market, characterised by low levels of purchasing and house price deflation. It appeared plausible that such an atypically severe recession would impact on the societies, whose core business remained dependent upon housing finance, in ways not entirely captured by changes in the output variable employed here. Accordingly, we included a dummy variable [90s *RECESS*] for the years 1990–3 inclusive.[12] The results are shown in columns (3), (4) and (6). It can be seen that while this variable does attract a negative – though not significant – coefficient, our underlying results are substantially unaffected. In particular, the ATM coefficients are materially unaffected although the third lag for the non-intensive users is now significant.

6 Conclusions

Nowhere has the productive contribution of IT investment been subject to such scepticism as in the service sector, and in financial services in particular. However, as Griliches (1994) and others have pointed out, service sector data is particularly suspect, raising the possibility that an absence of observed productivity gains is attributable to measurement problems. This chapter has tried to circumvent many of the usual problems encountered in using service output and IT capital input data by concentrating upon the deployment of a homogeneous embodied IT application in a narrowly defined financial market. The paper has adopted the financial intermediation view of banking output as the level of real earning assets. It then examined the impact of ATM deployment on productivity and labour demand using an unbalanced panel of ninety-three UK building societies between 1981 and 1993.

In contrast with some of the more pessimistic general assessments of the consequences of IT investment in service industries, our results are consistent with a large productivity effect. The labour productivity comparison is indicative of substantial gains to adopters both immediately following the introduction of an ATM system and subsequently. We then used a standard dynamic labour demand model, augmented by ATM variables to capture the contemporaneous and lagged effects of the innovation. The model appears well determined, with wage, output and lagged employment coefficients of appropriate signs and highly plausible magnitudes, whilst the regression diagnostics are satisfactory. The contemporaneous and subsequent, up to three lags, effects of ATM adoption on labour demand are negative and significant. While the initial introductory shock is relatively small, the long- run elasticity was equivalent to a *ceteris paribus* 57 per cent fall in labour demand for adopters. When the adopters were dichotomised into intensive and non-intensive adopters it was clear that the former displayed a substantially greater fall in demand than the latter, but even the non-intensive adopters showed a big decline relative to non-adopters. Furthermore, given that an absence of priors on the specifics of the post-adoption effects led us to a somewhat ad hoc treatment of the ATM variables, it is reassuring to find that the basic substantial, significant negative result is very robust to specification changes.

These results give no support to the IT productivity paradox view and tend to confirm the most recent findings of Lichtenberg (1995) and Brynjolfsson and Hitt (1996) on US manufacturing data that, at least in its later applications, IT investment may be associated with large productivity gains. However, it is suspected that our results represent upper bound estimates and may exaggerate the true employment impact of ATM introduction for a number of reasons:

First, the introduction of an ATM system is likely to be associated with the deployment of other subsidiary IT inputs whose separate effects we are unable to observe here.

Second, in an industry emerging from a highly regulated and protected state it is likely that managerial inputs are of mixed quality. Under such circumstances it is not implausible to expect that the more innovative managers, as judged by their earlier adoption of ATM technology, are also the more successful in adjusting to their changing circumstances. In a related context, Geroski, Machin and Van Reenan (1993) have demonstrated that innovative firms have superior performance even prior to the very innovation which is used to identify such firms.

Third, as with the adoption of any new technology, it would be expected that during the diffusion stage the early adopters will tend to be among those potential users who have most to gain from the use of the technology. Therefore, the gains to the adopters may exaggerate the realisable gains to other potential users.

And finally, in common with other IT investments, ATM introduction appears to raise output via the demand-side effects from improved service quality. To some extent these output gains will represent transfers from non-adopters which might be expected to diminish in size as the diffusion process proceeds.

One additional caveat concerns the quality of labour inputs. This chapter has made the usual simplifying assumption of homogeneous labour inputs. However, it has been widely conjectured that IT adoption impacts significantly upon the types of labour required. In the context of the building society sector, which makes extensive and growing use of part-time employees, further research is needed to determine the role of IT adoption on substitutability between different types of labour.

Notes

1 The authors would like to express their appreciation to Ray Barrell, Geoff Mason and Mary O'Mahony, editors of this volume, the participants at the ESRC/NIESR Conference on 'Productivity and Competition' including Gavin Cameron, Zvi Griliches, Bronwyn Hall and Paul Stoneman and to Sourafel Girma, Hilary Ingham, Peter Wright and Mike Wright for helpful comments on earlier drafts and to the Leverhulme Trust for financial support.
2 There is an extensive literature on hedonic price measurement in computers – see for example, Griliches (1988), Siegel and Griliches (1992) and Jordenson and Stiroh (1995) for discussions of the problems.
3 Macdonald and Lamberton (1983) address these issues using a more qualitative approach.
4 It also appears likely that the productive impact of IT-intensive technology varies considerably across industries: see Kwon and Stoneman (1995).
5 The Association for Payment Clearing Service (APACS) Yearbook (various years) provides data on the total numbers of UK ATMs, for banks and building societies, including data on utilisation. Regressions of the logarithms of the total

number of ATM cards and the number of withdrawals per machine, respectively, yielded the following results for 1981–91:

log ATM cards = 10.21 (139.8) + 0.077 (6.54) time
$R^2 = 0.82$, t-statistics in parentheses, and
log withdrawals per machine = 0.76 (1268.8) + 0.04 (29.5) time
$R^2 = 0.99$.

These growth rates indicate a rapid increase in number of potential users over the period and a substantial increase in the utilisation rates for each installed machine.

6 Diffusion within the building societies sector occurred rather later than in the commercial banks: see APACS Yearbooks (various years).

7 Four years after initial adoption three societies have ATM/branch ratios greater than unity and another six greater than 0.33. Nine adopters have ATM/branch ratios less than 0.2.

8 For example, the mean difference of the productivity of the non-adopters and subsequent adopters in years $t-3, t-2$ and $t-1$ (t-statistics in parentheses) was 0.024 (0.523), 0.075 (1.541) and 0.165 (2.888) with the latter alone significant. This line of reasoning is consistent with Bernard and Jensen's (1995) findings that US exporting firms' productivity differs from that of their non-exporting rivals, and indeed with the finding of Geroski, Machin and Van Reenan, (1993) that innovative firms differ from the rest even before the innovation that distinguishes them. We are grateful to Gavin Cameron for this comparison.

9 Even were the data to be available, construction of a satisfactory proxy for this variable would be difficult. Some inputs presumably have a uniform rental cost across the sample whilst others (for example, office rental) probably correlate with the cross-sectional variation in wages. However, the only data that are available for the entire sample and period are those in the annual accounts. These do contain asset data, but the use of such to derive capital cost measures is obviously unsatisfactory since the intermediation approach uses assets as the output measure. Drake (1995) has tried to proxy capital cost with a weighted combination of management expenses and depreciation charges. However, the correspondence between such a measure and the economic concept of a rental cost of capital is unclear. Drake recognises this and provides alternative weights (from one to zero) for each component. Given all these difficulties it was considered preferable to follow the Murray and White (1980) lead and assume an across sample constant cost of capital.

10 The introduction of higher order lags became problematic because the panel length had been shortened by first differencing and instrumentation.

11 We experimented with including a pre-adoption year dummy, reasoning that prior to start-up an ATM system would require additional labour. In the event this coefficient was totally insignificant.

12 While the intermediation approach to bank production views output as the level of earning assets, there are clearly multiple functional activities required to turn deposits into housing loans. Some of these functions were presumably severely under-utilised during the housing market recession with possible implications

for the desired level of employment. Furthermore, the book value of commercial assets may become an unreliable output measure during a period characterised by a high level of mortgage repossessions and house price deflation.

References

Arellano, M. and Bond, S. (1991), 'Some tests of specification for panel data: Monte Carlo evidence and an application to employment equations', *Review of Economic Studies*, 58, pp. 277–97.

Bailey, M.N. and Gordon, R.J. (1988), 'The productivity slowdown, measurement issues and the explosion of computer power', *Brookings Papers on Economic Activity*, 2, pp. 347–432, Washington DC, The Brookings Institution.

Barua, A., Kriebel, C. and Mukhopadhyay, T. (1991), 'Information technology and business value', University of Texas at Austin Working Paper.

Bernard, A.B. and Jensen, J.B. (1995), 'Exporters, jobs and wages in US manufacturing, 1976–1987', Washington DC, Brookings Papers on Economic Activity: Microeconomics, pp. 67–112.

Berndt E.R. and Malone, T.W. (1995), 'Information technology and the productivity paradox: getting the questions right', *Economics of Innovation and New Technology*, 3, pp. 177–82.

Berndt, E.R. and Morrison, C. (1994), 'High tech capital formation and economic performance in US manufacturing industries', *Journal of Econometrics*, 65, pp. 9–43.

Brynjolfsson, E. (1993), 'The productivity paradox of information technology', *Communications of the ACM*, 35 (December), pp. 66–77.

Brynjolfsson, E. and Hitt, L. (1996), 'Paradox lost? Firm-level evidence on the returns to information systems spending', *Management Science*, 42, 4, pp. 541–55.

Brynjolfsson, E., Malone, T.W., Gurbaxani, V. and Kambil, A. (1994), 'Does information technology lead to smaller firms?' *Management Science*, 40, 12, pp. 1628–44.

Daniels, K.N. and Murphy, N.B. (1994), 'The impact of technological change on household transaction account balances', *Journal of Financial Services Research*, pp. 113–19.

David, P.A. (1990), 'The dynamo and the computer: an historical perspective on the modern productivity paradox', *American Economic Review*, 80, 2, pp. 355–61.

Drake, L. (1995), 'Testing for expense preference behaviour: the UK building societies', *Service Industries Journal*, 15, pp. 50–65.

Geroski, P.A., Machin, S. and Van Reenan, J. (1993), 'The profitability of innovating firms', *Rand Journal of Economics*, 24, pp. 198–211.

Griliches, Z. (1988), *Technology, Education and Productivity*, Oxford and New York, Blackwell.
(1994) 'Productivity, R&D and the data constraint', *American Economic Review*, 84, pp. 1–23.
(1999), 'R&D and productivity growth: recent evidence and the uncertain future', paper given to Conference on Productivity and Competitiveness, National Institute of Economic and Social Research.
Hamermesh, D.S. (1993), *Labour Demand*, Princeton, Princeton University Press.
Hannan, T.H. and McDowell, J.M. (1984), 'The determinants of technology adoption: the case of the banking firm', *Rand Journal of Economics*, 15, 3, pp. 328–35.
Hardwick, P, and Ashton, J. (1997), 'Estimating inefficiencies in depository institutions: a survey', Bournemouth University Discussion Paper No. 9.
Ingham, H. and Thompson, S. (1993), 'The adoption of new technology in financial services: the case of building societies', *Economics of Innovation and New Technology*, 2, pp. 263–74.
Jorgenson D. and Stiroh, K. (1995), 'Computers and growth', *Economics of Innovation and New Technology*, 3, pp. 295–333.
Kwon, M.J. and Stoneman, P. (1995), 'The impact of technology adoption on firm production', *Economics of Innovation and New Technology*, 3, pp. 219–33.
Lichtenberg, F.R. (1995), 'The output contributions of computer equipment and personnel: a firm-level analysis', *Economics of Innovation and New Technology*, 3, 3–4, pp. 201–17.
Loveman, G.W. (1994), 'An assessment of the productivity impact of information technology', in Allen, T.J. and Scott Morton, M.S. (eds), *Information Technology and the Corporation of the 1990s: Research Studies*, Cambridge Mass., MIT Press.
Macdonald, S. and Lamberton, D. (1983), 'Tradition in transition: technological change and employment in Australian trading banks', *Australian Computer Journal*, 15, 4, pp. 128–39.
Malone, T.W. and Rockart, J.F. (1991), 'Computers, networks and the corporation', *Scientific American*, 265, 3, pp. 128–36.
Meyer-Krahmer, F. (1995), 'The effects of new technology on employment', *Economics of Innovation and New Technology*, 2, pp. 131–49.
Morrison, C.J. (1997), 'Assessing the productivity of information technology equipment in US manufacturing industries', *Review of Economics and Statistics*, 79, pp. 471–81.
Morrison, C.J. and Berndt, E.R. (1990), 'Assessing the technology of information technology equipment in US manufacturing industries', NBER Working Paper No. 3582, January.
Murray, R.D. and White, R.W. (1980), 'Economies of scale and deposit taking institutions in Canada: a study of British Columbia Credit Unions', *Journal of Money, Credit and Banking*, 12, pp. 58–70.
Nickell, S. (1981), 'Biases in dynamic models with fixed effects', *Econometrica*,

49, pp. 1417–26.

Osterman, P. (1986), 'The impact of computers on the employment of clerks and managers', *Industrial and Labour Relations Review*, 39, pp. 175–86.

Roach, S.S. (1987), *America's Technology Dilemma: A Profile of the Information Economy*, New York, Morgan Stanley.

—— (1991), 'Services under siege: the restructuring imperative', *Harvard Business Review*, September–October, pp. 82–9.

—— (1992), *Technology Imperative*, New York, Morgan Stanley.

Salop, S.C. (1990), 'Deregulating self-regulated shared ATM networks', *Economics of Innovation and New Technology*, 1, 3, pp. 85–96.

Sealey, C.W. and Lindley, J.T. (1977), 'Inputs, outputs and a theory of production and cost at depository financial institutions', *Journal of Finance*, 32, pp. 1252–63.

Siegel, D. and Griliches, Z. (1992), 'Purchased services, outsourcing, competencies and productivity in manufacturing', in Griliches, Z. (ed.), *Output Measurement in the Service Sector*, Chicago, University of Chicago Press.

Tushman, M. and Nadler, D. (1986), 'Organizing for innovation', *Californian Management Review*, 28, pp. 74–92.

Wilson, D.D. (1995), 'IT investment and its productivity effects: an organisational sociologist's perspective on directions for future research', *Economics of Innovation and New Technology*, 3, 3–4.

5 Productivity and service quality in banking: commercial lending in Britain, the United States and Germany

GEOFF MASON, BRENT KELTNER AND KARIN WAGNER[1]

1 Introduction

Market service industries account for ever-increasing shares of economic activity in advanced industrialised nations. In the United States, for instance, such industries now represent some 42–43 per cent of total output and employment and in Britain and Germany the respective figures are 38–40 per cent and 35–36 per cent (O'Mahony, Oulton and Vass, 1998). Yet while the overall prosperity of advanced economies is increasingly dependent on the productivity and efficiency of market service providers, there have been few detailed investigations of international differences in service sector productivity performance (in contrast to the smaller but better-researched manufacturing industries).

As a contribution towards filling this gap, this chapter reports on a detailed comparison of productivity and service quality in commercial bank lending in the United States, Britain and Germany. Our findings are based on data collected from matched samples of banking establishments in the three countries which are all engaged in lending to 'middle market' or 'midcorporate' business customers. This bank service area was chosen because this size of customer – typically firms with approximately $1 million to $300 million in annual sales – is frequently neglected in studies of bank credit provision to industry even though the success of middle market enterprises is arguably a key factor in national economic performance (Coopers and Lybrand, 1994).

The chapter is ordered as follows: in section 2, we describe the research methodology employed and the samples of bank lending offices in each country. Section 3 reports on detailed comparisons of labour productivity levels in the three samples of bank offices. Section 4 considers the

117

available evidence on the quality of service provided by middle market lenders in each sample. In section 5 we report measures of relative financial performance and compare the intensity of bank–customer relationships in each country. Section 6 examines the links between relative productivity performance and the quality and utilisation of physical and human capital inputs. Section 7 concludes.

2 Methodology and sample selection

The central advantage of cross-country comparisons of matched samples of production or service establishments is that inter-country differences in productivity performance can be related to detailed information about physical capital inputs, workforce skills and training, work organisation and business strategy in each sample. However, particular care needs to be taken to ensure that the samples are not only well-matched for product/service area and mix of employment-sizes but are also adequately representative of national populations of establishments in respect of these and other key criteria. At the same time, the matched-establishment methodology has obvious limitations arising from its focus on narrowly-defined product/service areas and the heavy costs of field work which necessitate reliance on smaller samples of establishments, and shorter periods of time in each establishment, than might be possible with alternative methods of enquiry. Hence in the present study we take care throughout to compare our sample-based findings against published information pertaining to the commercial banking industries as a whole in the three countries.

Research visits were made to a total of 50 different bank offices serving mid-corporate customers in the three countries (17 offices each in the US and Germany and 16 in Britain). These visits were carried out between June 1995 and December 1996. In order to capture some of the regional diversity in economic conditions and customer bases within each country, the visits were clustered in several different regions: the West Coast, Mid West and New England in the US; the North West, North East and South East in Britain; and Berlin, Baden-Württemberg, and Nordrhein-Westfalen in Germany.

As table 5.1 shows, the median size of bank office in the three samples was broadly similar (25–30 people) although there were fewer offices employing fewer than ten people in the US than in Britain or Germany. In other key respects, however, we allowed sharp differences in the composition of the three samples to emerge in order to reflect inter-country differences in the range of banking institutions that are active in commer-

Table 5.1 *Employment size-distribution of national samples of bank lending offices*

	Britain	US	Germany
Employment size-group:			
Under 10	3	1	4
11–20	4	7	6
21–30	5	3	3
Over 30	4	6	4
Total	16	17	17
Median size of office	29	30	25

Note: The median size is that where half of all employees are in offices above that size and half below it.

cial lending. Thus, in order to achieve a representative mix of American banks serving this market, the US sample was divided fairly evenly between national, regional and local banks, the majority of which were 'full service' banks (that is, engaged in both retail and commercial activities). In addition, a small minority of locally-active wholesale banks engaged solely in commercial lending were also included in the US sample. In Britain just under two thirds of sample offices were part of large national banks and the remainder were medium-sized banks whose networks are mainly confined to their 'home' regions. The German sample comprised a fairly even mix of privately-owned banks with a national focus and regionally-active public savings and co-operative banks (table 5.2A).

We also encountered considerable diversity between the three countries in the extent to which bank lending offices were targeting different sizes of customer, and this too is reflected in the respective samples. In general, the highest levels of customer 'segmentation' were found in the US and the American sample included five out of seventeen offices catering solely for small mid-corporate customers (typically with annual sales in the range $1–25 million) and another five offices dealing with larger mid-corporate customers (with annual sales ranging from $25 million up to $300 million). Only seven of the US bank offices served a full range of mid-corporate (and sometimes larger) customers compared with 13 out of 16 British banks catering to a wide range of customers from the same office. In Germany the extent of segmentation was also considerably greater than in Britain but less than in the US (table 5.2B).

The lending offices were all identified from commercial listings of bank establishments in each country and preliminary telephone interviews were

Table 5.2 *Distribution of national samples of bank lending offices by geographical coverage of parent banks and extent of customer segmentation*

	Britain	US	Germany
(A) Geographical coverage of parent banks			
National	10	5	9
Regional	6	7	8
Local	0	5	0
(B) Target customer range of sample offices[a]			
Small mid-corporate	0	5	5
Upper mid-corporate	3	5	3
Broad mid-corporate	3	5	5
Full range			
Mid-corporate and larger customers	10	2	4
Total	16	17	17

Notes:
[a]Bank offices classified as follows: Small mid-corporate: most customers in sales range from $1m to $25m. Broad mid-corporate: most customers in sales range from $5m to $150m. Upper mid-corporate: most customers in sales range from $25m to $300m. Full range: broad client base of mid-corporate and larger customers in sales range from $1–20m to $300m-plus.
[b]Customer sales data converted to US$ at PPP exchange rates for 1996: US $1.00 = £0.67 = DM 2.18.

carried out with senior managers before formal permission for a visit was sought. In the course of visits data were gathered through a combination of semi-structured face-to-face interviews with managers and written questionnaires which managers were asked to complete after the visit. The interviews were typically conducted with senior managers, who were sometimes accompanied by middle-ranking lending managers. In a number of the banks supplementary interviews were also conducted with human resource managers. At least two researchers from different countries participated in each interview. The purpose of the written questionnaires was to collect detailed information on bank lending output, revenue sources and the labour inputs involved in preparing business credit proposals (see section 3.2 for further details of these survey instruments). Following the research visits intensive follow-up enquiries by telephone were made with participating banks in each country in order to discuss the interpretation of data and other issues of interest.

3 Comparisons of labour productivity levels in mid-corporate lending

3.1 Measurement issues and previous cross-country comparisons

The conceptual and practical difficulties involved in measuring banking output and productivity levels have been widely discussed. Three distinctively different approaches to defining bank output which emerge from the literature are:
1. the 'national accounts approach' which seeks to measure value added as reflected in the profits and income arising from employment in banking (including profits derived from net interest receipts as well as from bank fees and charges);[2]
2. the 'production approach' which regards banks as using physical and human capital inputs to produce different types of financial service;[3]
3. the 'intermediation approach' which treats banks as intermediators between lenders and borrowers rather than as producers of financial services.[4]

In general, there is agreement that each of these approaches has some merit depending upon the purposes of the research in question and the types of data which are available for analysis (see, for example, the debates between Berger and Humphrey and discussants in Griliches, 1992, and the summary article by Colwell and Davis, 1992).

One recent cross-country study of comparative labour productivity levels in banking based on the production approach was carried out by the McKinsey Global Institute (1992). This focused on retail (high street) banking and used physical output data such as the number of payment transactions per year, the number of deposit accounts held by customers and the stock of retail credits outstanding to measure bank output. Their estimates for 1989 found average labour productivity levels in American retail banking to be approximately 47 per cent higher than in Germany and 56 per cent higher than in Britain. These differences were attributed in the main to greater usage of automation and information technologies and more efficient use of labour in the US industry – apparently reflecting greater competitive pressures in the US retail banking market than in the two European countries.

By contrast, O'Mahony, Oulton and Vass (1998) use national accounts data covering a much wider range of banking and finance services to derive estimates of value added per hour worked in 1993 which show average labour productivity levels in the US and German industries to be fairly similar, and only about 10 per cent above productivity levels in Britain. The differences with the McKinsey findings for retail banking illustrate the sensitivity of all such calculations to differences in methodology and in industrial coverage.

3.2 *Comparisons of lending output per employee*

For the present study focusing on mid-corporate lending – and with a particular interest in identifying the links between relative productivity performance, service quality and physical and human capital inputs – we adopted a production approach to the measurement of lending output and associated labour inputs.

In this area of banking the lending process usually begins with one or more meetings between representatives of a bank and representatives of the firm requesting a loan. If these meetings end with continued interest on the part of both parties, the bank then begins the process of formally evaluating the prospective borrower. This evaluation may include some or all of the following elements: a review of the client company's financial accounts, a 'peer-group comparison' of its performance against other similar enterprises, an industry and market evaluation, an on-site audit, and an assessment of its senior managers' experience and competence. If the bank loan officer (or 'relationship manager') concerned forms a view that the credit request should be granted, the company evaluation is then written up into a formal lending proposal which is submitted for a final credit decision to a higher level of decision-making authority within the bank.

In our survey of individual mid-corporate offices in the three national samples, each office was asked to provide information on the number of new business loans which were completed (i.e. accepted by customers) within a recent 12-month period (including new loans supplied to existing customers as well as new customers). Data were also requested on the total money value of such new lending, the different sources of net income earned by each office over the same period, the labour inputs associated with different forms of bank service and the nature and intensity of bank–customer relationships. In addition, further information on the labour inputs associated with each phase of the client evaluation and loan decision-making process was sought by asking individual banks to complete detailed questionnaires based on two hypothetical business credit requests (described in Section 3.3). These survey instruments were prepared in the light of advice from participating bankers in each country. In total, fully completed lending surveys were returned by half of the fifty bank offices in our three country samples; some twenty-eight different bank offices completed at least one case study questionnaire.[5]

Table 5.3 *Descriptive statistics of national samples of mid-corporate lending offices*

	Britain	US	Germany	Tests for equality of respective sample means (*p*-values)		
				Britain/ US	Britain/ Germany	US/ Germany
Average number of completed loans per employee over 12-month period[a]	3.9 (1.1)	6.0 (2.5)	15.4 (3.3)	0.472	0.009 ***	0.029 **
Average size of completed loan (US$)[b]	708000 (127400)	1161000 (661500)	373000 (129500)	0.588	0.291	0.188
Average proportion of net income deriving from lending-related activity (%)	47 (6.2)	68 (6.2)	71 (4.4)	0.021 **	0.007 ***	0.661
Average proportion of formal loan proposals prepared by bank staff on which bank willing to proceed (%)	86 (2.9)	88 (5.4)	98 (1.2)	0.786	0.0007 ***	0.067 *
Average proportion of loan proposals authorised by bank which are accepted by customers (%)	91 (3.5)	77 (4.8)	93 (2.8)	0.037 **	0.686	0.0101 **
No. of observations	8	10	7			

Source: Lending Survey.
Notes: Standard errors in brackets. F-tests of equality of sample means: *** = Inter-country differences statistically significant at 1 per cent level or better; ** = 5 per cent level; * = 10 per cent level.
[a] Refers to total direct plus indirect employment in bank lending offices (including secretarial and clerical support staff as well as relationship managers and credit analysts) without adjustment for labour inputs apparently devoted to non-lending activities (see table 5.5, note b for details of such an adjustment).
[b] Converted to US$ at 1996 PPP exchange rates: US $1.00 = £0.67 = DM 2.18.

Preliminary analysis of the data suggested that the average number of completed loans per employee in the German sample far exceeded that in either the US or Britain (table 5.3, row 1). However, when the average size of loan in each country was converted to a common currency (US$) using purchasing power parity (PPP) exchange rates for 1996, the average loan size in Germany was found to be only a third as high as the average US loan size and only half the average British loan size (table 5.3, row 1).[6] (Although there was considerable variation in average loan sizes within each country sample – and the standard errors attached to these estimates are therefore relatively high – the broad pattern of inter-country difference appears to reflect a greater tendency for German customers to split their borrowing between different banks than occurs in Britain or the US; see section 5.2 below for further discussion of this point.)

For some relatively simple types of banking transaction, the use of physical measures of output can be seen as avoiding the problem described by Smith (1989) as 'digit illusion', for example, the error involved in assuming that a cheque for $100,000 is as time consuming to process as ten cheques for $10,000. However, in the context of mid-corporate business lending, all information gathered in the course of this study suggests that the labour inputs associated with preparing a loan proposal are strongly positively related to the size and complexity of the credit request, and hence that simple productivity comparisons based on the average number of completed loans per employee need to be adjusted for inter-country differences in average size of loan.[7]

In order to derive a suitable adjustment factor for this purpose, we carried out a regression analysis of the data supplied by 25 banks (10 US, 8 British, 7 German) where, taking natural logarithms, the dependent variable 'ln output' referred to the average number of completed loans by each bank office and the independent variables were defined as follows:

ln loansize = average size of completed loan in US$ (converted at 1996 PPP exchange rates)

ln lendshare = proportion of total net income deriving from lending activity

ln emp = total employment in each bank office (typically comprising senior manager, relationship managers, credit analysts and clerical and secretarial staff).

The lending-related share of total net income in the analysis was included as a proxy measure of the labour inputs devoted to lending as opposed to other forms of revenue generation in the bank offices. Interest and fee income from lending represented much the same share (around 70 per cent) of total net income in the US and German samples but only 47 per cent in

Britain (table 5.3, row 3): this disparity is consistent with evidence on the relatively large share of income deriving from charges on customers in the British banking system as a whole but it also reflects higher levels of resources devoted to non-lending deals such as management buy-outs in the British sample.[8]

The results of this analysis are summarised in table 5.4. The negative (and statistically significant) coefficients on ln loansize in equations (1) and (3) show that, at a given level of employment, there is indeed a clear inverse relationship between the number of loans completed and the average size of loans. The coefficient on ln loansize also remains stable in size, sign and significance in equations (4) and (5) which control for 'country-specific' factors affecting productivity as well. By contrast, the coefficients on ln lendshare in equations (2), (3) and (4) are unstable and not statistically significant; in subsequent analysis we report sensitivity tests on all assumptions made using revenue data of this kind.

Table 5.5, row 1 shows that if no allowance is made for differences in average loan size, then – as noted above – the average annual number of completed loans per employee in German bank offices is substantially higher than in either the US or Britain. A rough adjustment for inter-country differences in lending-related labour inputs sharply improves the British position relative to both other countries but does not disturb the German–US differential (table 5.5, row 1a). However, the coefficient on ln loansize in equation (5) suggests that, at a given level of employment, a doubling of the average size of loan is associated with a reduction in the number of loans (and hence in labour productivity) of $2^{-0.67} = 0.63$, i.e. 37 per cent. When this adjustment factor is used to standardise for average loan size in the three countries, then average lending output per employee in the German sample is estimated to be some 15 per cent higher than in the US and almost two thirds higher than in Britain (table 5.5, row 1b). This country ranking conforms with that indicated by the regression results in table 5.4, equations (4) and (5).[9] Further adjustments for inter-country differences in average annual hours worked per employee increase the estimated German productivity lead over the US to 23 per cent but do not greatly affect the German–British differential (table 5.5, row 1c).

As an alternative to a physical measure of lending output (number of loans), productivity comparisons could have been based on a financial measure, namely, the total money value of lending by each bank office over a given time period. On the basis of survey data supplied by participating banks in each country, the average money value of lending per employee (converted to a common currency) was estimated to be about 20 per cent higher in the US sample than in Germany and over twice the British level

Table 5.4 *Regressions of annual loan output on average size of loan, lending-related shares of net income and bank office employment*

	Dependent variable: ln output (total number of completed loans)				
	(1)	(2)	(3)	(4)	(5)
Constant	7.25	−1.09	6.71	6.32	6.68
	(6.37)***	(−0.40)	(2.86)***	(4.97)***	(5.31)***
ln loansize	−0.72	–	−0.72	−0.70	−0.67
	(−5.61)***		(−5.28)***	(−4.33)***	(−4.25)***
ln lendshare	–	0.65	0.12	−0.71	–
		(1.01)	(0.26)	(−1.22)	
ln emp	0.75	1.04	0.75	0.67	0.68
	(4.16)***	(3.97)***	(4.07)***	(3.62)***	(3.81)***
Country dummies:					
United States	–	–	–	0.70	0.39
				(1.54)	(1.07)
Germany	–	–	–	1.18	0.74
				(2.18)**	(1.78)*
Adjusted R^2	0.74	0.39	0.73	0.77	0.75
SEE	0.74	1.13	0.76	0.72	0.72
No. of observations	25	25	25	25	25

Notes: *t*-statistics in brackets. Statistically significant at: *** = the 1 per cent level or better; ** = 5 per cent level; * = 10 per cent level. Country dummies set to zero for bank offices located in Britain.

(table 5.5, row 2). However, this comparison also needs to be adjusted for inter-country differences in the average size of loan, in this case to take account of the fixed, 'start-up' element of labour input associated with processing a new credit request whatever its size. And since the average size of loan for each bank office equates to total money value of lending divided by the number of loans, the resulting productivity estimates are identical to those derived earlier using the number of loans (adjusted for average loan size) as an output measure (table 5.5, row 2c).

Several further issues need to be discussed in relation to the definitions of both lending output and associated labour inputs which underlie these estimates.

Firstly, we have defined lending output in terms of *completed* loans, that is, formal loan proposals which have been both cleared by each bank's central credit authority *and* accepted by the customers for whom they were intended. As our survey data show that larger proportions of loan proposals tend to fail at each of these hurdles in the US than in either Britain

Table 5.5 *Estimated labour productivity levels in national samples of mid-corporate lending offices (Index numbers: US = 100)*

	Britain	US	Germany
1. Average annual number of completed loans per employee[a]	65	100	257
Sequential adjustments for inter-country differences in:			
(a) ratio of lending-related to non-lending labour inputs[b]	97	100	246
(b) average loan size[c]	69	100	115
(c) average annual hours worked per employee[d]	75	100	123
2. Average annual money value of lending per employee (US$)	40	100	83
Sequential adjustments for inter-country differences in:			
(a) ratio of lending-related to non-lending labour inputs[b]	59	100	79
(b) average loan size[e]	69	100	115
(c) average annual hours worked per employee[d]	75	100	123

Notes:
[a] Index numbers derived from table 5.3, row 1.
[b] Estimates based on assumption that inter-country differences in lending-related labour inputs are proportional to inter-country differences in the share of total net income derived from lending activity (see text for details of sensitivity test on this assumption). 'Net income' is here defined as: all income from loans, deposits, fees and services *less* interest expenses on loan financing and deposits measured before bad loan provisions and taxes.
[c] The adjustment factor used to standardise for average loan sizes in the three countries is derived from the regression results for equation (5) in table 5.4 which imply that, at a given level of employment, a doubling of the average size of loan is associated with a reduction in lending output of $2^{-0.67} = 0.63$, i.e. 37 per cent less.
[d] Based on estimates by O'Mahony, Oulton and Vass (1998) of average annual hours worked per employee in the banking and finance industries Index numbers (US = 100): Britain 93, Germany 94.
[e] Loan size adjustment factor also derived from the regression results for equation (5) in table 5.4 which imply that, if total money value of lending (= number of loans*average loan size) is taken as the output measure, then at a given level of employment, a doubling of the average size of loan is associated with an increase in total lending value of $2^{0.33} = 1.26$, i.e. 26 per cent more.

or Germany (table 5.3, rows 3 and 4), an alternative definition of loan output in terms of prepared proposals would improve estimated US productivity relative to the two European countries. However, our preferred definition (completed loans) is closer to the concept of 'gross output' which is widely used in productivity analysis.[10] As will be argued below, the finding that more labour-time put into preparing formal loan proposals is effectively 'lost' in the US than in the other two countries – seemingly owing to higher levels of competition for customers – is intrinsically interesting and needs to be taken into consideration in explaining the identified pattern of relative productivity performance in this sector of bank lending.

Secondly, in this industry where relationship managers are expected to maximise income earned through 'cross-selling' of different financial products and services to those bank customers for whom they are responsible, it is difficult to measure the labour inputs directly associated with lending. As described in table 5.5, note (a), our central productivity estimate is based on the assumption that inter-country differences in lending-related labour inputs are proportional to inter-country differences in the share of total net revenue derived from lending activity. However, the earlier regression analysis using these data did not identify any systematic relationship between lending productivity and lending-related shares of net income. The distribution of net income may well reflect other factors besides the mix of lending and non-lending activities, for example, inter-country differences in the degree of competition inhibiting the imposition of fees and charges for different services. Hence, we carried out a sensitivity test on our assumption of proportionality between revenues and labour inputs: if inter-country differences in lending-related labour inputs are instead assumed to be only half as great as those implied by the mix of lending and non-lending revenues in each country, then the estimated German–US productivity differential is found to change hardly at all but average lending productivity in the British sample would fall to just under 60 per cent of the US level and less than half the German level.[11]

Thirdly, managerial estimates supplied during interviews suggested that, on average, US relationship managers spend (or are expected to spend) substantially more time on seeking to 'develop' or 'acquire' new customers than their German or British counterparts: an estimated 45 per cent of American relationship managers' time is spent in search of new business rather than in serving existing customers (compared with 15 per cent and 20 per cent, respectively, in Germany and Britain).

However, rather than adjust our productivity estimates for inter-country variation in 'non-productive' labour-time, we go on to argue that such differences in the approach to new business development reflect important differences in market conditions and in business strategies in each country, and thus are better considered as potential explanatory factors in any assessment of relative performance.

3.3 Case studies of mid-corporate lending

Our central productivity estimates based on survey data were broadly supported by data collected from bank offices on two different case studies of business credit requests. These credit scenarios were based on actual (anonymised) case details supplied by participating bankers in the study and respondents were asked to estimate the total employee-hours directly associated with three distinct phases of the lending process: the initial contacts with the client company; preparation of a formal client company evaluation; and the approval or denial of a formal loan proposal. Case 1 involved a relatively small credit request for $330,000 while Case 2 was based on a much larger request for a $3 million revolving credit plus a $5 million term loan. A majority of US respondents divided between one set of bank offices for which Case 1 was relevant and another set which would typically deal only with larger (Case 2-type) customers. This contrasted sharply with the British respondents which all dealt with both types of case and also with the German sample in which the respondents typically dealt either with both sizes of case or with the smaller (Case 1) end of the market alone.

For Case 1 the average total labour inputs in the US and German samples were estimated to be broadly similar (17–18 employee-hours), with British banks requiring an average 22–23 hours to process this size of loan. However, for Case 2 US banks serving the size of customer concerned reported an average labour input more than twice as large as that in Germany and over 50 per cent greater than in Britain (table 5.6). These initially surprising findings were confirmed in follow-up discussions with the banks concerned and we discuss their implications at length below. The Case 1 results suggest that US bank offices match the productivity performance of their German counterparts in servicing credit requests at the 'smaller' end of the mid-corporate market, with the British sample lagging in third place. By contrast, the Case 2 results suggest that American banks serve larger customers in a much more labour-intensive way than either German or British banks. Although these results are based on very small sample sizes, and the standard errors attached to the estimates are thus relatively high, they are broadly consistent with the pattern of inter-country productivity difference shown in table 5.5. In particular, the lower measured

Table 5.6 *Average direct employee-hours associated with case studies of business loan requests: Case 1: US$330,000 credit request; Case 2: US$3 million revolving credit and a $5 million term loan*

	Britain	US	Germany	Tests for equality of respective sample means (p-values)		
				Britain/ US	Britain/ Germany	US/ Germany
Average total employee-hours required						
Case 1	22.6 (4.3)	17.8 (3.2)	17.2 (1.6)	0.419	0.210	0.826
No. of observations	10	7	10			
Case 2	39.3 (7.6)	62.1 (14.4)	26.7 (3.0)	0.117	0.117	0.019 *
No. of observations	10	8	8			
Average lending output per employee-hour (standardised loan size):		*Index numbers (US = 100)*				
Case 1	79	100	104			
Case 2	158	100	233			

Source: Case study questionnaires returned by lending offices.
Notes: Standard errors in brackets. F-tests of equality of sample means: * = Inter-country differences statistically significant at 10 per cent level.

productivity in US banks serving the upper end of the mid-corporate market helps to explain the overall US productivity shortfall relative to Germany which has been identified. In the next section we go on to assess the relationship between measured productivity performance and the quality of lending services provided in each country.

4 Service quality in middle market lending

Cross-country comparisons of bank lending productivity based on physical measures of loan output may not do justice to systematic differences

Table 5.7 *Measures of 'problem lending' and in national samples of mid-corporate lending offices (by percentage)*

	Britain	US	Germany
Proportion of total outstanding lending which in last 3 years has been:			
put under special scrutiny (e.g. watch or oversight list)	3.6	5.5	6.3
moved to 'special assets' or 'workout'[a]	1.1	1.8	3.1
written off or charged off	1.7	0.6	2.5
No. of observations	7	11	7

Source: Lending Survey.

in the *quality* of the credit services in question. In particular, if banks give more priority to detailed examination of business loan requests than to providing customers with a quick answer, then it is possible that the credit arrangements which ensue are better tailored to the specific needs of individual clients and may also be less likely to result in customer failure to repay the loan. These issues deserve full consideration in any evaluation of bank performance.

4.1 Lending 'failure' rates

From the banks' perspective, the quality of lending is best measured in terms of loan failure rates. In our survey of bank offices, we obtained information on three different categories of 'problem lending' in the three years prior to our visits (1992–5):

- loans put under 'special scrutiny' (e.g. placed on a watch or oversight list)
- loans moved to 'special assets' or 'workout' departments (with specialist bank staff working closely with customers to maximise the chances of loan recovery)
- loans which have been written off or 'charged off'

As table 5.7 shows, at this time the mid-corporate lending offices in our German sample had larger proportions of current lending in all three problem categories than either their US or British counterparts. This is at first sight surprising given the widespread belief that German banks typically have closer relationships with their customers than do American and

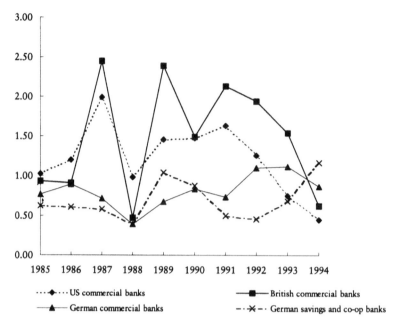

Figure 5.1 *Provisions (net) as a proportion of total loans in US, British and German commercial banks and German savings and co-operative banks, 1985–94*
Source: OECD (1996).
Note: Provisions generally include, in part or in full, charges for value adjustments in respect of loans, credits and securities, book gains for such adjustments, losses on loans and transfers to and from reserves for possible losses on such assets.

British banks and are thus better equipped to assess the risks attached to new credit requests.[12] However, data for the wider commercial banking industries over a ten-year period show that loan failure rates are heavily cyclical in nature. In general, between 1985 and 1994 the ratio of net provisions for bad debts and other adjustments to total lending in Germany was both lower and less volatile than in Britain or the US. But towards the end of this period – the time to which our sample data refer – the German economy was experiencing its most severe recession since the Second World War and bad debt provisions in German banking did rise above the levels then prevailing in the American and British industries (figure 5.1).

These aggregate data on bad debt provisions and other value adjustments suggest that no firm conclusions about the average quality of credit risk assessment procedures in each country can be drawn from sample data which apply to a relatively short period of time. However, data derived from the two lending case studies enable us to assess certain aspects of the quality of business lending *as experienced by borrowers*, in particular, the speed with which decisions on new loan requests are made and the extent to which banks seek to tailor credit services to the specific needs of individual business customers.

4.2 Speed of response to credit requests

In the smaller of the two cases, referring to a loan request for some $330,000, the average time required for a full appraisal to be carried out and a decision given to the customer was roughly the same in the US and German samples (five working days) and about 50 per cent longer in Britain (table 5.8, part A). This pattern of response times largely corresponds with the estimated differences in average labour productivity levels for this size of case (table 5.6). In the second case study (a request for $3 million revolving credit and a $5 million term loan), the three-country ranking in average response times is also the same as in the labour productivity comparison based on this case: the average time taken to 'turn around' a credit request of this size and complexity is just over 10 working days in the German sample, 11.5 days in Britain and 22 days in the US (table 5.9, part B). However, the difference in average response times between German and British banks is much smaller than the estimated productivity gap between the two samples.

4.3 Quality of analysis in credit appraisal

In order to assess the extent to which these differences in efficiency and speed of response reflect variations in the quality of analysis applied to the credit appraisal process, participating bank offices were asked to list (in order of importance) the five main types of information which they would seek from clients in order to guide their decision-making. They were also asked to cite the main types of information missing from the case study which they would need to obtain.

In all three countries a high priority is naturally given to basic accounts information (for example, client company balance sheets and audited management accounts). However, in terms of other information requirements, the Case 1 results show clear differences between US banks, on the one hand, and British and German banks on the other. In general most US banks catering for this 'lower end' of the mid-corporate market are

Table 5.8 *Indicators of service quality associated with case studies of business loan requests: Case 1: US$330,000 credit request*[a]

Employment-weighted averages	Britain	US	Germany
	No. of working days		
A. Average time required for final decision on credit request to be given to customer:	7.5	5.0	5.3
B. Main types of information sought from client in order to evaluate credit request:	*Rank ordering of importance (1=highest)*[b]		
1. Basic accounts information	2	1	1
2. Senior managers' experience	3	6=	6
3. Industry/market information	4	6=	3
4. Business prospects	1	4	2
5. Personal finances of client firm's owners	7	2	4
6. Collateral	6	3	5
7. Client firm's previous banking record	5	5	7
Spearman rank correlation coefficients on ordering of importance of loan evaluation criteria:			
Britain/US: –0.125; Britain/Germany: 0.536; US/Germany: 0.482.			
C. Main types of information used during evaluation of credit request:	*Percentage of bank offices*[c]		
Types of information entered into standardised computer program used for credit evaluation:			
Financial information supplied by client firm	100	100	100
Information on client firm's management	70	20	50
Information from industry projection/market analysis	70	60	50
Other types of information	50	20	65
No. of observations	10	6	9

Source: Case study questionnaires returned by lending offices.
Notes:
[a]Involves a five-year-old, family-owned wholesale distributor of roofing materials, with total sales last year of just under $4 million. The company is a potential new customer at the bank and has made a request for $330,000 in credit to finance its ongoing business expansion.
[b]Ranking based on proportions of banks in each sample listing five most important types of information sought from customers, weighted by order of priority indicated by respondents.
[c]Percentages rounded to nearest five.

more concerned to obtain information relating to collateral and loan security (for example, information on the personal finances of the client firm's owners) than they are to invest time in analysing data necessary for a good understanding of the client's business and industry (table 5.8, part B).

By contrast, while British and German banks do naturally seek information relating to collateral and security, they report an even greater interest in obtaining data on business prospects (for example, projections of market demand) and industrial structure. Indeed, in the British sample the types of information necessary for 'business understanding' were ranked as more important even than basic accounts information: this appeared to reflect comments made by many British lending managers in interviews that a great deal of 'security-based' lending had failed during the early 1990s recession in Britain and that they were now trying harder to develop and maintain long-term relationships with clients.

In each country it is now commonplace for bank lending staff to enter relevant data into standardised computer programs designed to facilitate a credit risk assessment and guide the final recommendation on the credit request. In so doing use is invariably made of other information sources apart from the client company itself and a majority of banks in each country (including the US) reported that information based on some kind of industry projection or market analysis would be entered into the computer-based assessment. However, as table 5.8, part C shows, the range of information taken into account by US banks for this size of credit request is typically narrower than in either British or German banks and the final credit decisions in US banks are likely to be primarily based on analysis of client financial data.

Our case study evidence therefore suggests that at the 'low end' of the mid-corporate market, US banks place more emphasis on pushing through relatively high volumes of security-based lending than on analysing each client's business and industry in any depth. By contrast, German banks manage to achieve a similar level of efficiency as US banks in dealing with this size of credit request while seeking to base their decisions on a more detailed understanding of clients' businesses and credit needs. In this they are greatly helped by the ability of most German customers to provide banks with up to date sales, cost and cash flow data in a form which includes detailed peer-group comparisons of those clients' performance.[13] For the size of credit request signified by Case 1, British banks now appear to be belatedly aiming to build relationships with their customers similar to those in Germany but most of their staff still lack the experience that German bank employees have in rapidly accessing and analysing the information necessary for a detailed credit evaluation.

Table 5.9 *Indicators of service quality associated with case studies of business loan requests: Case 2: US$3 million revolving credit and a $5 million term loan*[a]

Employment-weighted averages	Britain	US	Germany
	No. of working days		
A. Average time required for final decision on credit request to be given to customer:	11.5	21.9	10.3
B. Main types of information sought from client in order to evaluate credit request:	*Rank ordering of importance (1=highest)*[b]		
1. Basic accounts information	2	1	1
2. Senior managers' experience	3	2	6
3. Industry/market information	4	5	3
4. Business prospects	1	3	2
5. Personal finances of client firm's owners	7	6	5
6. Collateral	5	4	4
7. Client firm's previous banking record	6	7	7

Spearman rank correlation coefficients on ordering of importance of loan evaluation criteria:

Britain/US: 0.821**; Britain/Germany: 0.679; US/Germany: 0.607.
[** = statistically significant at 5 per cent level.]

	Britain	US	Germany
C. Main types of information used during evaluation of credit request:	*Percentage of bank offices*[c]		
Types of information entered into standardised computer program used for credit evaluation:			
Financial information supplied by client firm	100	100	100
Information on client firm's management	70	100	60
Information from industry projection/market analysis	70	90	40
Other types of information	50	65	60
No. of observations	10	8	8

Source: Case study questionnaires returned by lending offices.
Notes:
[a]Involves a twenty-year-old manufacturing company which makes sporting equipment. The company was founded in 1974 as a manufacturer of various sporting goods and began designing and manufacturing goods to the higher-end sporting industry in 1989. Total sales last year were $54 million. It is operated and owned by its original owners. It has made a request for a £3 million revolving credit and a $5 million term loan.
[b]Ranking based on proportions of banks in each sample listing five most important types of information sought from customers, weighted by order of priority indicated by respondents.
[c]Percentages rounded to nearest five.

However, if we consider the much larger credit request (Case 2) at the 'high end' of the mid-corporate market, a very different pattern of difference between the three national samples starts to emerge. Recall that the American bank offices targeting this section of the market devoted on average more than twice the labour inputs required by German banks to deal with this size of credit request, and US labour inputs were also more than 50 per cent higher than in Britain. Table 5.9, part B shows that this investment of labour time and effort by US banks is reflected in a very different ranking of information needs than applied in Case 1: in contrast to those American bank offices which focus on relatively small mid-corporate clients, the US offices dealing with larger customers attach at least as much importance to information about clients' business prospects, industry position and management experience as they do to information related to security and collateral.[14]

Similarly, if we consider the types of information which are entered into computer-assisted assessments of risk and creditworthiness (table 5.9, part C), the US banks' responses for Case 2 suggest that their analysis is far more thorough and comprehensive than for Case 1, and may also be even more detailed than the analysis typically carried out in German and British banks. By contrast, the responses of the two sets of European banks for Case 2 suggest that, although this larger case is inherently more complex and time-consuming than Case 1, their basic strategy of seeking to base credit decisions on business understanding as well as loan security is much the same for all sizes of credit request in the mid-corporate market.

The limitations of our case study data must be acknowledged: the banks' responses relate to hypothetical situations and shed more light on the breadth of information incorporated into credit assessments in the three countries than on the depth of analysis carried out. However, our findings lend some support to the following propositions regarding lending productivity and service quality:

- German leadership in lending productivity over the US and Britain does not seem to be achieved at the expense of providing a lower quality of service (as guaged in terms of speed of response to customers and effort to gain a detailed understanding of clients' credit needs).
- The poor productivity performance of British banks relative to US banks partly reflects the fact that, at the lower end of the mid-corporate market, British banks seek to provide a more in-depth analysis of client credit needs than do US banks targeting this section of the market. However, the British–German productivity gap seems to reflect differences in lending efficiency rather than in service quality.

- Estimates of average lending productivity levels in the US sample conceal large variations in the quality of service provided to different market segments. In general, high productivity in the volume-oriented low end of the market is offset by a very labour-intensive approach to servicing credit requests from larger mid-corporate clients.

5 Productivity and financial performance

5.1 Comparisons of net income per employee-hour

Although the strategic decisions taken by US banks to devote considerable resources to servicing larger clients (and to new business development in general) undermine US performance in terms of measured labour productivity, they appear to be soundly based in terms of financial performance: in respect of revenue generation, average net income per employee-hour in the US was found to be some 14 per cent higher than in Britain and almost 60 per cent above the German level.[15] This ranking of the three samples in terms of revenue generation is consistent with data for the wider commercial banking industries, which show net income as a proportion of total assets to be consistently lower in German banks than in either their British or American counterparts over the period 1985–94 (Figure 5.2).

5.2 Revenue generation and intensity of bank–customer relationships

In order to gain a better understanding of the relationship between productivity and financial performance, we sought additional information from sample banks on the longevity of their client base and the extent to which their clients depended on them alone for credit and other banking services. The results, summarised in tables 5.10 and 5.11, highlight several interesting points of contrast between the three samples and help to explain why German leadership in terms of labour productivity does not translate into higher levels of financial performance relative to the US and Britain.

In all three countries a majority of customers have been with their current banks for five years or more. The proportion of customers in this category is greatest in Germany (71 per cent) and lowest in the US (60 per cent) but these differences in longevity are neither large nor statistically significant (table 5.10). Far greater differences can be observed in the

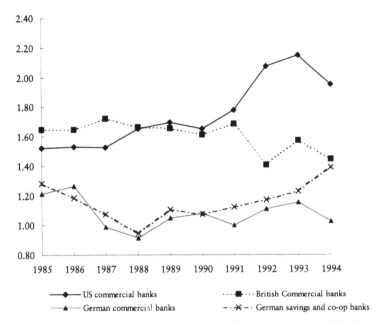

Figure 5.2 *Net income as a proportion of total assets in US, British and German commercial banks and German savings and co-operative banks, 1985–94*
Source: OECD (1996).
Note: 'Net income' is here defined as 'Gross income less operating expenses'. Operating expenses primarily include salaries and other employee benefits; expenses for property and equipment and related depreciation expenses; and taxes other than income or corporate taxes.

extent to which bank customers consolidate all their product holdings with a single bank in the three countries.

For example, the proportion of clients borrowing two-thirds or more of their working capital from the same bank is significantly lower in the German sample (32 per cent) than in Britain (63 per cent) or the US (74 per cent) (table 5.11). A similar pattern of difference – also statistically significant – is found in the proportions of clients using the same bank for two-thirds of more of their transactions and transmission services (i.e. cheque and transfer payments). These findings are consistent with other evidence of German business customers' tendency to split their financial service business between different banks (Pfeiffer, 1986; Edwards and Fischer, 1994).[16]

Table 5.10 *Duration of bank–customer relationships in national samples of mid-corporate lending offices (by percentage)*

Employment-weighted averages	Britain	US	Germany	US Small mid-corporate clients	US Large mid-corporate clients
Proportion of existing business clients which have been with bank:					
less than 2 years	14	14	9	15	13
2 to 5 years	20	26	20	32	24
more than 5 years	66	60	71	53	63
Total	100	100	100	100	100
No. of observations	7	11	10	5	6

Chi-square tests of equality of sample distributions (p-values):
US/Britain 0.586, US/Germany 0.247, Britain/Germany 0.530,
US (small)/US (large) 0.342.

Source: Lending Survey.

Taken together with the labour- and time-intensive strategy adopted by US banks towards larger mid-corporate customers, and the recent efforts by British banks to deepen their understanding of clients' business and market prospects, these findings suggest that (in this sector of banking at least) the common stereotype of 'relationship banking' in Germany contrasting with 'volume banking' in the US and Britain does not apply in any straightforward way.

Table 5.11 also suggests that the average proportion of clients using other bank services such as investment or deposit products is lower in Germany than in the US or Britain. This variation between the three samples in the extent of 'cross-selling' of different products may help to explain why the relatively strong German performance in lending productivity coexists with weaker performance in terms of financial measures. Conversely, the higher level of cross-selling in the British sample as compared to Germany presumably helps to compensate in financial terms for the lower levels of efficiency in lending operations in British banks.

The highest overall levels of product cross-selling were found in the US sample which had the highest average net income per employee-hour. As table 5.11 shows, the bulk of this cross-selling is carried out (as expected) by those US banks which focus on providing relatively high-quality, labour-intensive services to more profitable customers in the upper segment

Productivity and service quality in banking

Table 5.11 Business customers' use of bank services in national samples of mid-corporate lending offices (by percentage)

Employment-weighted averages	Britain	US	Germany	US small mid-corp. clients	US large mid-corp. clients	F-tests for equality of respective sample means (p-values):			
						Britain /US	Britain /Germany	US/ Germany	US (small)/ US (large)
Proportion of existing business clients:									
borrowing two-thirds or more of working capital from bank	63 (11.2)	74 (6.2)	32 (7.8)	82 (19.6)	70 (19.8)	0.331	0.034 **	0.0004 ***	0.297
using bank for two-thirds or more of transaction or transmission services	72 (12.0)	80 (7.1)	41 (10.6)	80 (8.2)	80 (9.1)	0.517	0.062 *	0.004 ***	0.999
using deposit or investment products provided by bank in last 12 months	51 (5.7)	60 (8.3)	46 (17.4)	33 (13.9)	71 (14.7)	0.377	0.788	0.404	0.065 *
No. of observations	7	11	7	5	6				

Source: Lending Survey.
Notes: Standard errors in brackets. Inter-country differences statistically significant at: *** = 1 per cent level or better; ** = 5 per cent level; * = 10 per cent level.

of the mid-corporate market. By contrast, in US banks oriented towards volume lending to smaller business customers, the extent of product cross-selling is actually lower than in either of the two European countries.

6 Relative productivity, work organisation and capital inputs

German banks come under strong pressure to economise on labour and achieve high levels of efficiency in their lending operations from the twin forces of relatively high operating costs and competitive pressures in a relatively 'crowded' banking market. Although the German commercial banking industry is much less concentrated than the US industry, private commercial banks in Germany have to compete with large numbers of public savings banks and co-operative banks for both commercial lending and retail business (Keltner, 1995). All these different kinds of bank have outlets in almost every town and city regardless of their size, making it rare for there to be fewer than five or six banks competing in any local banking market.

By contrast, in Britain the dominance of commercial banking by a handful of large nationwide banks is well-documented and the relatively strong performance by British sample offices on net income per employee – in contrast to their weaker performance on productivity – is consistent with evidence presented earlier on the relatively high proportion of British lending office income deriving from charges on customers and other types of non-lending activity (table 5.3). In the US, commercial banks serving midcorporate customers face at least as much competition in local and regional banking markets as their German counterparts but (as described above) their strategic response is to economise on labour costs only at the lower end of the middle market while seeking to maximise earnings through cultivation of larger, potentially more profitable customers at the upper end of the market.

As described in section 4.3, German lending productivity is enhanced by bank offices' swift access to and long experience of using nationwide business databases to support peer-group comparisons and market analyses when processing new credit requests. The German sample also had the lowest proportion of 'failed' lending proposals, that is formally-prepared proposals which were either not cleared by the banks concerned or were not accepted by the customers for whom they were intended (section 3.2). And German relationship managers devote a smaller proportion of their time to searching for prospective new clients than do their American or British counterparts (section 3.2).

In terms of physical capital inputs, the levels of usage of IT equipment for credit appraisal in the three samples were broadly similar. US lending offices began working with personal computers (PCs) earlier than in the two European countries and have therefore had longer to adapt their working methods and procedures to the potential efficiency benefits offered by this equipment. Among other things, this showed up in a greater degree of 'streamlining' in work organisation in US offices; for example, some relationship managers work independently with only occasional recourse to back-up staff (Mason, Keltner and Wagner, 1998). However, as described above, relatively high productivity in the volume-oriented low end of the market in the US is offset by the very labour-intensive approach to servicing credit requests from larger mid-corporate clients.

While US banks were more economical in the use of credit analysts to support relationship managers than are British or German banks, the German sample had the lowest proportion of clerical/secretarial staff. This was associated with a high proportion of German clerical employees who have undergone a full apprentice training after leaving school (in contrast to their generally less well-trained American and British counterparts). In all three samples relationship managers and credit analysts had similarly high levels of formal qualifications and training although the specific formation of these groups of employees reflected pronounced inter-country differences in education and training systems and labour markets.[17]

The relatively poor productivity performance of British banks partly reflects the greater use of support staff in Britain than occurs in either the US or Germany. At the same time British banks still lack the equivalent of the large databases available in Germany to support credit appraisal based on a detailed understanding of clients' businesses and markets. And relative to the US, British productivity performance is further diminished by the fact that, at the lower end of the market, British banks seek to provide a more in-depth analysis of clients' credit needs than do US banks targetting this size of customer (Section 4.3).

7 Summary and assessment

Our comparison of labour productivity performance in bank lending to mid-corporate business customers in the United States, Britain and Germany has led to the following main findings:

1. After standardising for average loan sizes in the three countries, average lending output per employee-hour in the German sample is estimated to be some 23 per cent higher than in the US and almost two-thirds higher than in Britain.

2. German leadership in lending productivity does not seem to be achieved at the expense of providing a lower quality of service than in the US or Britain (as gauged in terms of speed of response to customers and efforts to gain a detailed understanding of clients' credit needs). Rather German banks are under considerable pressure from relatively high labour costs to economise on staff numbers (and on the time spent on credit appraisals by relationship managers and credit analysts) even while seeking to maintain service quality.

3. The poor productivity performance of British banks relative to US banks partly reflects the fact that, at the lower end of the mid-corporate market, British banks seek to provide a more in-depth analysis of client credit needs than do US banks targeting this section of the

market. However, the British–German productivity gap seems to reflect differences in efficiency in processing business credit requests rather than in service quality.

4. Within the US sample there was considerable variation in the quality of service provided to different market segments. In general, high productivity in the volume-oriented low end of the market is offset by a very labour-intensive approach to servicing credit requests from larger mid-corporate clients which is designed to maximise sales of a wide range of banking services to the potentially most profitable customers in the mid-corporate size bracket.

5. This strategy of market segmentation is much more developed in the US than in the two European countries and is associated with higher average net income per employee-hour in the US than in either of the two European countries. The British sample is ranked second ahead of Germany in this respect, and the overall findings are consistent with data on financial performance for the wider commercial banking industries in recent years.

The study has highlighted several measurement issues which affect the estimates of relative productivity levels in mid-corporate lending. For example, we have defined loan output in terms of completed loans rather than the number of proposals which were formally prepared by lending staff, and the labour inputs associated with lending have been defined to include time spent in efforts to develop new customer contacts as well as time devoted to servicing existing contacts.

Thus the factors enhancing German productivity performance relative to the US and Britain include both its relatively small proportion of 'failed' loan proposals (that is, proposals which are formally prepared but fail to receive authorisation within the bank or acceptance by customers) and a larger proportion of relationship managers' time spent on serving existing clients.

Conversely, in spite of the many advantages which the US sample derives from its more innovative forms of work organisation and economical use of support staff, its measured productivity performance is diminished by several factors:

- a higher proportion of time lost in preparing formal proposals which are not accepted by the customers for whom they were intended –

often due to price competition from other lending institutions.
- the larger proportion of time which American relationship managers devote to searching for prospective new clients.
- the deliberately more labour-intensive strategy adopted by banks serving the higher end of the market to help maximise revenues from selling a full range of products (and to stave off competition from rivals).

These findings shed light on the difficulties in interpreting inter-country comparisons of banking productivity based on published data sets which cannot distinguish between labour inputs directly devoted to the production of bank service outputs and other labour inputs which are absorbed in responding to competitive market pressures. Although the business strategies adopted by US mid-corporate lending offices apparently contribute to higher average net income per employee-hour in the US than in either of the two European countries, they are also associated with greater expenditures of American lending managers' time on activities which are essentially 'unproductive' in terms of their contribution to bank output.

Notes

1 We are glad to acknowledge financial support for this enquiry which was provided by the Sloan Foundation; the Foundation is not responsible in any way for the views expressed in this paper. We are grateful to David Finegold for detailed comments and suggestions, to Kurt Pertsch for excellent research assistance and to Ray Barrell, Mary O'Mahony and Nick Oulton and participants at the 1998 ESRC/NIESR Conference on Productivity and Competitiveness in London for comments on earlier versions of this chapter. Responsibility for errors is ours alone.
2 See O'Mahony, Oulton and Vass (1998) for further discussion of this approach.
3 This is the term favoured by Colwell and Davis (1992). A production approach is similar to that described by Berger and Humphrey (1992) as a 'value added' approach with 'important' banking outputs defined as those requiring large expenditures on labour and physical capital.
4 Berger and Humphrey (1992) prefer to describe this as an 'asset approach' to defining bank output.
5 Although we would have liked a higher response rate, we acknowledge that the detailed information requested was more time-consuming for some banks to provide than we had anticipated.
6 Since no international statistical authority produces data on purchasing power parities for financial services, we follow O'Mahony, Oulton and Vass (1998) in using PPP exchange rates for private consumption recalculated to exclude non-

146 Productivity, innovation and economic performance

market services. For 1996 this procedure yields the following PPP exchange rates: US $1.00 = £0.67 = DM 2.18.

7   There are several other reasons apart from complexity why larger business loan requests typically absorb more labour-time in their assessment than smaller loan requests: not only are there potentially greater losses involved (in the event of bad debts), but larger customers are also viewed as potentially more profitable to the bank office, hence more worthy of time-intensive personal contacts and relationship-building.

8   Between 1985 and 1994 non-interest income accounted for an average 39 per cent of gross income in British commercial banks compared to 32 per cent in US commercial banks, 31 per cent in German commercial banks, 19 per cent in German co-operative banks and 15 per cent in German savings banks (OECD, 1996). For discussion of trends in income sources and the 'thoroughgoing attitude to implementing charges' in British banking, see Bank of England (1993) and Robbie and de Hoest (1992).

9   Note that in both equations (4) and (5) in table 5.4, where the UK is taken as the base country variable, the positive coefficient on the German country dummy variable is statistically significant but the positive coefficient on the US dummy is poorly determined owing to a high degree of variation around the mean productivity level in the US sample. As described in section 3.3, this reflects a marked divergence in measured productivity performance between US banks serving different sections of the mid-corporate market.

10 'Gross output' in manufacturing is typically defined as: Total sales and work done *less* any increase in work in progress and goods on hand for sale. Business loan proposals which are formally prepared but not brought to fruition may sometimes later be re-usable (after further work and updating), but in most cases the time spent on preparing them is non-recoverable.

11 More precisely, in these circumstances average annual lending output per employee-hour (with a standardised loan size) would be (index numbers): US 100, Germany 126, Britain 58.

12 See Davis (1994) and Vitols (1995) for a discussion of these issues.

13 In particular, German clients can ask their tax accountants to prepare detailed statements of their sales performance, cash flow, interest payable, capital depreciation, materials, labour and other costs and other variables in a form which is directly comparable with financial data for groups of other similar firms recorded on the nationwide DATEV database.

14 Note that the Spearman rank correlation coefficients on the ordering of importance of loan evaluation criteria for US banks in relation to British and German banks are markedly higher for the larger Case 2 credit request (table 5.9, part B) than for the relatively small Case 1 (table 5.8, part B).

15 Average annual net income per employee was estimated as follows (index numbers): US 100, Britain 88, Germany 63. 'Net income' is here defined as: all income from loans, deposits, fees and services *less* interest expenses on loan financing and deposits measured before bad loan provisions and taxes (converted to US$ at 1996 PPP exchange rates: US $1.00 = £0.67 = DM 2.18).

16 In a recent survey of 1,127 German enterprises with 500 or fewer employees, Harhoff and Körting (1998) found that the concentration of borrowing decreases strongly with firm size. The firms in their sample had a median number of two different borrowing relationships with financial institutions, rising to three for firms with 100–500 employees (*ibid.*, table 4). This compares with a median number of one lending institution per business customer in a late 1980s survey of 3,404 US businesses in the under-500 employees size-bracket (Petersen and Rajan, 1995, table 1). Harhoff and Körting also found that German firms with 100–500 employees still typically met about two thirds of their total credit needs from one institution, which suggests that some caution needs to be attached to the estimate for our small sample of German lending offices in table 5.11, row 1. However, it should be noted that firms with 100–500 employees typically correspond only to the 'bottom half' of the mid-corporate market as it has been defined in the present study.

17 For further details of inter-country differences in bank employees' education and training, see Mason, Keltner and Wagner (1998, section 7).

References

Bank of England (1993), 'Bank lending to small businesses', *Bank of England Quarterly Bulletin*, 33, 1.

Berger, A. and Humphrey, D. (1992), 'Measurement and efficiency issues in commercial banking', in Griliches, Z. (ed.), *Output Measurement in the Service Sectors*, Chicago, University of Chicago Press.

Colwell, R. and Davis, E. (1992), 'Output, productivity and externalities: the case of banking', Bank of England Working Paper Series No. 3, August.

Coopers and Lybrand (1994), *Made in the UK: the Middle Market Survey*, London, Coopers and Lybrand.

Davis, E. (1994), 'Banking, corporate finance and monetary policy: an empirical perspective', *Oxford Review of Economic Policy*, 10, 4.

Edwards, J. and Fischer, K. (1994), *Banks, Finance and Investment in Germany*, Cambridge, Cambridge University Press.

Griliches, Z. (1992) (ed.), *Output Measurement in the Service Sectors*, Chicago, University of Chicago Press.

Harhoff, D. and Körting, T. (1998), 'Lending relationships in Germany – empirical evidence from survey data', Discussion Paper No. 98–21, Mannheim, Zentrum für Europäische Wirtschaftsforschung.

Keltner, B. (1995), 'Relationship banking and competitive advantage: evidence from the US and Germany', *California Management Review*, 37, pp. 45–72.

Mason, G., Keltner, B. and Wagner, K. (1998), 'Productivity, technology and skills in banking: commercial lending in the United States, Britain and Germany',

Paper presented to ESRC/NIESR Conference on Productivity and Competitiveness, London, February.

McKinsey Global Institute (1992), *Service Sector Productivity*, Washington, DC, McKinsey & Co.

OECD (1996), *Bank Profitability: Financial Statements of Banks, 1985–94*, Paris, OECD.

O'Mahony, M., Oulton, N. and Vass, J. (1998), 'Market services: productivity benchmarks for the UK', *Oxford Bulletin of Economics and Statistics*, 60, 4, pp. 529–51.

Petersen, M. and Rajan, R. (1995), 'The effect of credit market competition on lending relationships', *Quarterly Journal of Economics*, 110, pp. 407–43.

Pfeiffer, H. (1986), 'Grossbanken und Finanzgruppen: ausgewaehlte Ergebnisse einer Untersuchung der personellen Verflechtungen von Deutscher, Dresdner und Commerzbank', *WSI Mitteilungen*, July, pp. 473–80.

Robbie, M. and de Hoest, P. (1992), 'Bank profits in the 1990s', *Banking World*, April.

Smith, A. (1989), 'New measures of British service outputs', *National Institute Economic Review*, May.

Vitols, S. (1995), 'Are German banks different?', WZB Discussion Paper FS I 95–308, Berlin, November.

6 Productivity growth in an open economy: the experience of the UK

GAVIN CAMERON, JAMES PROUDMAN AND
STEPHEN REDDING[1]

1 Introduction

Between 1970 and 1990, value-added per worker in the UK grew at an average annual rate of 1.9 per cent.[2] However, this aggregate figure conceals considerable variation both across sectors and over time. Value-added per worker in services rose at an average annual rate of 0.5 per cent, while within manufacturing the corresponding figure was 3.0 per cent. In the sub-period from 1973–9 – corresponding to the peak-to-peak of the business cycle – the average annual rate of growth of value-added per worker in manufacturing was 1.7 per cent and in services 0.7 per cent. In contrast, in the second peak-to-peak business cycle period 1979–89, these figures rose to 3.7 per cent and 0.8 per cent respectively.

This differential growth performance across sectors and over time was matched by considerable differences in international openness, as measured variously by trade flows, trade barriers, capital flows and flows of ideas. For example, between 1970 and 1990, the ratio of exports to domestic output in manufacturing rose from 17.7 to 30.0 per cent. Within manufacturing, the average share of exports over the period 1970–90 ranged from 5.4 per cent in paper and printing to 79.8 per cent in computing.[3]

The objective here is to examine the extent to which the variation in rates of economic growth over time and across sectors in the UK is related to differences in the degree of international openness, where openness is defined as the magnitude of impediments to international flows of goods and services, factors of production and ideas.

The endogenous growth literature provides a theoretical framework within which the process determining long-run growth rates is represented either in terms of an increase in the variety or as a rise in the quality of the goods produced by an economy. Skilled labour may be employed either in the production of current output or in research and other activities associated with innovation. The rate of growth of output is determined by the rate of introduction of the new designs for goods discovered in the research sector. This itself is a function of the amount of skilled labour employed in research and the productivity of that research.

In a world with many economies at different stages of economic development, it is also likely that, over the medium-term, economies behind the technological frontier may grow more rapidly by investing in the adoption of technologies discovered in their more advanced counterparts.[4] International openness may therefore affect an economy's growth rate by influencing either the rate of innovation or the rate of adoption of existing technologies. Grossman and Helpman (1991), for example, examine the relationship between international openness and the rate of innovation in advanced economies.[5]

One way in which trade affects growth is by promoting increased specialisation.[6] If these sectoral shifts are biased towards fast-growing sectors then the economy's growth rate will rise. As we shall see in section 2, sectoral shifts appear to have played little role in the growth experience of UK manufacturing since 1970.

Having discounted this channel, the remainder of the chapter concentrates on the effects of trade within individual sectors. In this context, there are five interrelated channels through which international trade may affect rates of productivity growth:

1. Trade has a positive effect on growth because it may be directly responsible for the transfer of technology between countries with differing productivity levels. That is, trade may allow sectors to catch-up to productivity levels of more technologically advanced economies more quickly than otherwise. For example, trade may allow domestic firms to 'reverse engineer' products of their foreign rivals. This point concerns the adoption of ideas in production.
2. Trade has a positive effect on growth because it may be directly responsible for the spillover of ideas, thereby generating a larger pool of knowledge to assist future innovation, raising the productivity of research and boosting long-run growth rates. This idea relates to the dissemination of ideas among researchers.
3. Trade has a positive effect on growth because it eliminates incentives for duplication in innovation. The integration of countries' product markets through openness to trade places innovators in different

countries in competition with one another, and so gives them incentives to pursue ideas that are novel in the world economy. Trade thus tends to reduce duplication of research effort and hence increase the aggregate productivity of resources employed in innovation.
4. Trade has a positive effect on growth because it increases the market size available to successful researchers, increasing the incentive to engage in research.
5. Trade has an ambiguous effect on growth because it enhances the intensity of product market competition. Increased competition reduces the equilibrium profits to be derived from successful research, which in turn may either increase or decrease the incentive to engage in research.[7]

The structure of this chapter is as follows. Section 2 characterises the growth experience of the UK's manufacturing sectors and examines the effects of shifts of resources between sectors on productivity growth. Section 3 addresses the problem of moving from the conceptual definition of international openness (the degree of impediments to the free flow of goods and services, factors of production and ideas) to a number of quantitative measures of openness in all three of these dimensions. It then analyses the simple cross-section relationship between estimated rates of productivity growth and measures of international openness.

Section 4 presents a more formal econometric analysis. This is undertaken in the framework of a theoretical model within which an industry's productivity growth rate depends upon the difference between its initial level of productivity and the technological frontier. In this framework, international openness plays an important role in facilitating the transfer of technology from the frontier economy. We controlled for the effect of other potentially significant variables, such as the intensity of domestic research and development, differences in educational standards and changes in capacity utilisation. Using our resultss, we estimate implicit long-run levels of productivity in the UK relative to the US, and associate changes in these levels with variations in the main explanatory variables.

2 UK economic growth

A first step in the analysis is to characterise the nature of economic growth at the sectoral level.[8] The rate of growth of output can be decomposed into the contributions of increased hours worked, physical capital accumulation and a residual. This residual encompasses the effect of anything that influences the efficiency with which existing factors of production are used. It includes, for example, the influence of technology, unionisation, the extent of competition, training and factor hoarding. In practice, technical change is the dominant determinant of the residual – known as Total Factor Productivity (TFP) – which therefore provides a widely-used empirical measure of the rate of technological progress.

In discrete time, TFP growth may be approximated by the following Thörnqvist-Theil Divisia index,

$$\ln\left(\frac{A_j(t+1)}{A_j(t)}\right) = \ln\left(\frac{Y_j(t+1)}{Y_j(t)}\right) - (1-\tilde{\alpha}_j(t))\ln\left(\frac{K_j(t+1)}{K_j(t)}\right) \\ -\tilde{\alpha}_j(t)\ln\left(\frac{L_j(t+1)}{L_j(t)}\right) \quad (1)$$

where $\tilde{\alpha}_j(t) = \{\alpha_j(t) + \alpha_j(t+1)\}/2$. In this formulation, A represents TFP, Y is output (value-added), K is physical capital, L is labour input, and α_j is the share of payments to labour in value-added.[9]

While TFP constitutes a measure of efficiency with which both labour and physical capital are employed, it imposes greater theoretical restrictions on the data than a measure of labour productivity. The key assumptions here are perfect competition and constant returns to scale. In the remainder of this section, although we present results for TFP growth, the main results are robust to the use of either labour productivity or TFP as a measure of productivity. Furthermore, we replicated the analysis of this section for the whole economy without major changes to the results. For further detail see Cameron, Proudman and Redding (1997).

We make use of a detailed ONS data set, extended by Cameron (1996).[10] This covers manufacturing industry alone, disaggregated into nineteen subsectors. The decomposition of UK manufacturing output growth is shown in table 6.1.

Intuitively, productivity growth in the whole economy can be decomposed into that contribution made by productivity growth *within* individual sectors and that contribution made by switches in factor resources *between* sectors with differing levels of productivity.[11] The change in labour productivity may be expressed as the sum of within-sector productivity growth and changes between sectors in the share of hours worked,

$$\Delta\left(\frac{Y}{L}\right) = \sum_j \Delta\left(\frac{Y_j}{L_j}\right) w_j^L(t-1) + \sum_j \Delta w_j^L \frac{Y_j(t-1)}{L_j(t-1)} \quad (2)$$

Table 6.1 *Sources of output growth (value added) in UK manufacturing 1970–92 (percentage change p.a.)*

Sector	ISIC code	Output	Labour	Capital	TFP
Total	3	−0.18	−2.16	0.60	1.38
Food & drink	31	−0.23	−1.16	1.19	−0.26
Textiles	32	−1.49	−3.13	−0.12	1.76
Timber & furniture	33	−0.71	−1.84	0.86	0.27
Paper & printing	34	0.88	−1.43	0.99	1.32
Minerals	36	−2.33	−2.11	0.84	−1.06
Chemicals	35	1.40	−1.11	0.98	1.52
Chemicals nes[a]	351... 354–3522	−0.31	−1.62	0.82	1.10
Pharmaceuticals	3522	4.72	−0.65	1.52	3.85
Rubber & plastics	355+356	1.24	−1.21	0.87	1.58
Basic metals	37	−3.60	−5.43	0.09	1.73
Iron & steel	371	−4.20	−6.46	0.04	2.22
Non-ferrous metals	372	−1.93	−3.40	0.27	1.20
Fabricated metals	38	−0.01	−2.56	0.48	2.07
Metal goods	381	−1.01	−2.71	0.31	1.39
Machinery	382–3825	−1.54	−2.74	0.48	0.72
Computing	3825	7.62	−1.17	3.12	5.67
Other electrical eng.	383–3832	−0.31	−2.63	0.63	1.68
Electronics	3832	1.91	−2.28	1.18	3.01
Shipbuilding	3841	−4.14	−5.48	−0.06	1.40
Motor vehicles	3843	−1.22	−2.72	0.56	0.93
Aerospace	3845	2.58	−1.52	−0.07	4.17
Instruments	385	2.16	−1.67	0.88	2.95
Other manuf.	39	−1.38	−2.69	0.03	1.27

Note: [a]nes: not elsewhere specified.

where the weights w_j^L are each sector's share in total hours worked. The first term on the right-hand side of equation (2) is the 'within effect', and the second term is the 'between effect'. Following Bernard and Jones (1996a), a similar decomposition may be undertaken for TFP growth in aggregate manufacturing. Assuming that the production process in each manufacturing sector j is characterised by a common, time-invariant, Cobb–Douglas production technology, the decomposition is,[12]

$$\Delta TFP = \sum_j \Delta TFP_j w_j^{TFP}(t-1) + \sum_j \Delta w_j^{TFP} TFP_j(t-1) \quad (3)$$

Table 6.2 *Decomposition of productivity growth in the UK, 1970–92 (manufacturing and the whole economy)*

Shares of total growth o/a		Between	Within	Total
TFP	Whole economy	17.1	82.9	100.0
	Manufacturing	10.2	89.8	100.0
Labour productivity growth:	Whole economy	4.4	95.6	100.0
	Manufacturing	3.0	97.0	100.0

The results of these decompositions for both labour productivity and TFP growth in UK manufacturing (and the whole economy) are presented in table 6.2. As much as 97 per cent of labour productivity growth in manufacturing is explained by within-sector growth. The corresponding figure for manufacturing is somewhat smaller, 90 per cent, but still represents the majority of productivity growth.

Taken together, analysis of the productivity data suggests a number of stylised facts about the UK's growth performance:

1. Technological change was estimated to be the major source of output growth. Over the period from 1970 to 1992, manufacturing output fell (–0.18 per cent p.a.). Labour utilisation fell and the reduction in hours worked alone implied a –2.2 per cent rate of growth in output per annum. TFP and capital accumulation both made positive contributions to output growth, with the contribution of TFP (+1.38 per cent p.a.) substantially exceeding that of capital accumulation (+0.60 per cent p.a.).
2. There was nevertheless considerable variation in average growth rates of TFP (and labour productivity) across sectors. In manufacturing, these ranged from 5.67 per cent p.a. in computing, and over 3.5 per cent p.a. in pharmaceuticals, electronics and aerospace, to negative numbers in food and minerals. There was also considerable variation in the *level* of total factor productivity across sectors.
3. The share of output growth accounted for by TFP growth relative to that accounted for by capital accumulation was higher during the 1980s business cycle (1979–89) than during the 1970s cycle (1973–9).
4. There was an increase in the average growth rate of TFP (and labour productivity) in the 1980s compared with the 1970s. In manufacturing as a whole, TFP fell at an average annual rate of 1.88 per cent

p.a. between 1973 and 1979, but rose at an equivalent rate of 3.28 per cent between 1979 and 1989. An increase occurred in all manufacturing subsectors except computing.
5. The bulk of aggregate TFP growth and of labour productivity growth was generated by growth within sectors rather than by shifts in resources from low to high productivity sectors.

3 How open is UK manufacturing?

As suggested earlier, international openness can affect growth in a number of ways. The impact of most of these channels will be positive, although the effect of some is ambiguous. *In practice*, as will be seen below, there is considerable empirical evidence that the net effect in the UK is positive, and that international openness will be positively correlated with rates of economic growth.

Furthermore, it is plausible that a number of other economic factors will be important for explaining the observed pattern of economic growth. In particular, the introduction outlined how differences in the level of productivity compared with the US are likely to be a key feature of the interaction between openness and growth. Similarly, domestic rates of research and development, sectoral differences in educational standards, the degree of unionisation and changes in capacity utilisation are also frequently cited as important determinants of rates of productivity growth.[13]

In the light of this complexity, our approach to analysing the relationship between openness and growth is twofold. We begin by simply analysing partial correlations, and here the analysis draws upon Cameron, Proudman and Redding (1998a). This provides us with important stylised facts about the association between openness and growth.

As already noted, one problem with evaluating the relationship between openness and growth is that there are many different measures of international openness. We begin by combining the information contained in the different measures of international openness in order to characterise sectors as either open or closed. One way of doing this involves the use of the statistical technique of discriminant analysis. The objective of discriminant analysis is to select groups by emphasising both the similarities of the trade characteristics of the data within the same group *and* the differences between the representative properties of the groups.[14]

We used this technique to divide nineteen UK manufacturing sectors into groups labelled 'relatively open' and 'relatively closed' on the basis of five measures of openness: (exports/output (X/Y), imports/sales (M/S), inward

Table 6.3 *Period average growth characteristics of manufacturing industries classified as relatively closed in 1970*

Industry	M/S	X/Y	IFDI/Y	OFDI/Y	TWRD/Y	DTFP
Textiles	0.21	0.16	0.001	0.000	0.03	1.76
Timber & furniture	0.08	0.16	0.000	0.000	0.09	0.27
Minerals	0.11	0.06	0.000	0.000	0.06	−1.06
Iron & steel	0.12	0.07	0.000	0.019	0.08	2.22
Non-ferrous metals	0.21	0.37	0.000	0.057	0.19	1.20
Average closed	0.15	0.16	0.000	0.015	0.09	0.88

Table 6.4 *Period average growth characteristics of manufacturing industries classified as relatively open in 1970*

Industry	M/S	X/Y	IFDI/Y	OFDI/Y	TWRD/Y	DTFP
Food & drink	0.09	0.19	0.014	0.037	0.02	−0.26
Paper & printing	0.03	0.23	0.003	0.008	0.03	1.32
Chemicals	0.24	0.19	0.041	0.049	0.59	1.10
Pharmaceuticals	0.31	0.11	0.188	0.225	1.25	3.85
Rubber & plastics	0.13	0.06	0.019	0.000	0.20	1.58
Metal goods	0.12	0.09	0.016	0.031	0.09	1.39
Machinery	0.28	0.15	0.007	0.015	0.17	0.72
Computing	0.34	0.49	0.324	0.198	8.76	5.67
Other electrical eng.	0.19	0.24	0.066	0.041	2.56	1.68
Electronics	0.18	0.08	0.072	0.044	2.04	3.01
Motor vehicles	0.28	0.07	0.025	0.005	0.50	0.93
Aerospace	0.27	0.22	0.054	0.010	15.97	4.17
Instruments	0.35	0.29	0.285	0.174	2.75	2.95
Other manufacturing	0.32	0.19	0.076	0.245	0.86	1.27
Average open	0.22	0.19	0.085	0.077	2.56	2.10

foreign direct investment flows/output (IFDI/Y), outward foreign direct investment flows/output (OFDI/Y) and trade-weighted foreign R&D stocks/output (TWRD/Y)).[15] In order to try to address the endogeneity problem, we use the *1970* values of these variables to characterise sectors as relatively 'open' and 'closed'. We then seek to relate this characterisation based upon 1970 information to rates of economic growth after 1970. We present the results in tables 6.3 and 6.4.

The average values of the openness measures are considerably higher for the group of 'relatively open' sectors, than for the group of 'relatively

Table 6.5 *Cross-section variation in average TFP growth (1970–92) and openness across manufacturing sectors*

Openness measures (logs)	β (income/labour)	β (capital/labour)	β (TFP)
Exports/output (X/Y)	0.0109 **	0.0010	0.0112 **
(flow of goods)	(0.007)	(0.003)	(0.005)
Imports/sales (M/S)	0.0069	0.0025	0.0094
(flow of goods)	(0.007)	(0.003)	(0.006)
Inward FDI flows/			
output (IFDI/Y)	0.0026 **	0.0004	0.0023 **
(flow of capital)	(0.001)	(0.001)	(0.001)
Outward FDI flows/			
output (OFDI/Y)	0.0022 *	0.0005	0.002 **
(flow of capital)	(0.001)	(0.001)	(0.001)
Import weighted R&D/			
output (TWRD/Y)	0.0059 **	0.0004	0.0056 **
(flow of ideas)	(0.002)	(0.001)	(0.001)

Notes: ** indicates significance at the 95% level; * indicates significance at the 90 per cent level. Coefficients from least squares regression of productivity measures on initial (1970) measures of openness (standard errors in brackets)

closed'. At the same time, average productivity growth for the group of open sectors is found to equal 2.10 per cent p.a. compared with 0.88 per cent p.a. for the group of closed economies, suggesting a striking degree of association between openness and rates of growth of both TFP and output.[16] There is also a positive association between openness and *levels* of productivity.

Discriminant analysis offers a simple and easy to interpret way of illustrating the fact that sectors that are relatively open over a range of openness measures tend to be characterised by faster rates of productivity growth. The disadvantage of discriminant analysis is that it does not allow for differences in the degree of openness between members of the same group. The next step of the analysis therefore considers the association between openness and growth in a linear regression framework. That is, we regress the average annual rate of growth in labour productivity, TFP and the capital labour ratio from 1970–92 against the 1970 value of each measure of openness separately.

These cross-section regressions indicate that the ratios of inward FDI to output (IFDI/Y), outward FDI to output (OFDI/Y) and trade weighted R&D stocks to output (TWRD/Y) are significantly correlated with labour productivity growth. All of these measures, and the export to output ratio (X/Y) are significantly correlated with the rate of TFP growth. However, none of the measures of openness are significantly correlated with the capital/labour ratio. This is consistent with the hypothesis that openness affects growth through rates of technical change, not capital accumulation.

To address at least partly the endogeneity problem, the results reported in the main body of table 6.5 are coefficients derived using 1970 values of openness, estimated for nineteen sectors within manufacturing. These results are in fact fairly robust to alternative specifications. Similar results are derived (but not reported here) using an instrumental variables approach that addresses a different formulation of the endogeneity problem. In the instrumental variables approach, we regress the average annual rate of growth in TFP between 1970 and 1992 against the 1970 to 1992 period average values of each measure of openness separately, and use 1970 values of the same openness measure as an instrument.

The estimates for the nineteen manufacturing sectors in table 6.5 are also robust to the exclusion of extreme values. That is, it is possible to exclude the three sectors from the sample with the most extreme values of *either* growth rates *or* levels of openness without affecting the significance or sign of the results.

4 Productivity convergence and international openness

The empirical results presented in the preceding section provide evidence that openness is associated with growth across sectors: sectors that were relatively open in 1970 tended to have higher rates of productivity growth between 1970 and 1992.

However, this association reflects a partial correlation, which cannot be interpreted as a structural parameter, since no allowance has been made for interactions with and between other economic variables. In this section we therefore move on to consider the role of openness on growth in a more formal econometric framework derived from an underlying theoretical model.

The five channels identified earlier could have three proximate effects on rates of productivity growth in an economy behind the technological frontier. First, they may affect domestic rates of innovation. Second, they may affect the amount of technological knowledge that can be imported from the leading economy. Third, they may affect the rate at which this technology transfer occurs.

The basic theoretical framework is provided by Bernard and Jones

(1996a), extended to incorporate the effect of international openness, see Cameron, Proudman and Redding (1998b). Consider a world comprising two economies $i \in \{B, F\}$, each of which may produce any of a fixed number of manufacturing goods, $j = 1,...,n$. Each of these goods is produced with labour and physical capital according to a neoclassical production technology:

$$Y_{ij} = A_{ij} F_j(L_{ij}, K_{ij}) \qquad (4)$$

where K and L denote physical capital and labour respectively; and where A is an index of technical efficiency, which we define as Total Factor Productivity (TFP). The function $F(.,.)$ is assumed to be homogeneous of degree one and to exhibit diminishing marginal returns to the accumulation of each factor alone and is the same in both countries. A_{ij} may vary both across sectors and between economies.

At any point in time t and in any individual sector j, one of the two economies i will have a higher level of TFP than the other (except in the special case where TFP levels happen to be equal). This economy is termed the frontier economy F, while its counterpart is termed the backward economy B. In the present chapter we are concerned with the US and the UK. We find that UK TFP lies below US levels in all manufacturing sectors throughout the sample period (see below), and begin with the assumption that this will continue in the steady-state (an assumption that is supported by our parameter estimates).

Following Bernard and Jones (1996a), TFP in sector j of each economy i may potentially grow either as a result of sector-specific innovation or as a result of technology transfer from the frontier country,

$$\ln\left(\frac{A_{ij}(t)}{A_{ij}(t-1)}\right) = \gamma_{ij} + \lambda_j \ln\left(\frac{\omega_{ij} A_{Fj}(t-1)}{A_{ij}(t-1)}\right) \quad \gamma_{ij}, \lambda_j \geq 0, \ 0 > \omega_{ij} \leq 1 \qquad (5)$$

where γ_{ij} parameterises the rate of sector-specific innovation and ω_{ij} denotes the fraction of TFP in the frontier economy that may potentially be transferred to economy i, and the parameter λ_j characterises the rate at which technology transfer occurs.

If the economy i is the frontier economy, it already possesses the most advanced technologies and there is no potential for technology transfer (formally, $\omega_{Fj} = 1$ and $\ln(\omega_{ij} \cdot A_{Fj}/A_{ij}) = 0$). If economy i is behind the technology frontier, then it may benefit from technology transfer, although not all of the leading economy's technology may be relevant or transferable (formally, $0 > \omega_{Bj} \leq 1$ and $\ln(\omega_{ij} \cdot A_{Fj}/A_{ij}) > 0$).[17] Combining equation (5) for both the frontier and backward economies, one obtains a first-order difference equation for the evolution of TFP,

$$\ln\left(\frac{A_{Bj}(t)}{A_{Fj}(t)}\right) = (\gamma_{Bj} - \gamma_{Fj}) + \lambda_j . \ln \omega_{Bj} + (1-\lambda_j)\ln\left(\frac{A_{Bj}(t-1)}{A_{Fj}(t-1)}\right) \quad (6)$$

from which we may solve for the steady-state level of relative TFP ($\tilde{A} \equiv A_B / A_F$) in each sector j,

$$\ln \tilde{A}_j^* \equiv \ln\left(\frac{A_{Bj}^*}{A_{Fj}^*}\right) = \ln \omega_{Bj} + \frac{\gamma_{Bj} - \gamma_{Fj}}{\lambda_j} \quad (7)$$

where, for the initially backward economy to remain so in steady-state we require $\ln(A_{Bj}^* / A_{Fj}^*) < 0 \Leftrightarrow \gamma_{Fj} > \gamma_{Bj} + \lambda_j \ln \omega_{Bj}$. In the long run, the model implies that TFP in both economies grows at the same steady-state rate γ_{Fj} in sector j. The terms γ_{Bj}, λ_j, and ω_{Bj} determine the steady-state level of relative TFP in the backward economy and the rate of TFP growth in the transition to steady-state. For ease of exposition we can restate equation (5) in the following form:

$$\Delta \ln(A(t)) = \gamma_I + \lambda \ln\left(\frac{A^{US}(t-1)}{A(t-1)}\right) \quad (8)$$

where, $\lambda = f$ (openness, human capital, R&D, unionisation, etc)
$\gamma = g$ (openness, human.capital, R&D, unionisation, etc)

and where $A(t)$ and $\Delta\ln(A(t))$ denote the level and rate of growth of productivity in the UK respectively, and $A^{US}(t-1)$ denotes the level of productivity in the relevant sector in the US. We assume that $f(.)$ and $g(.)$ are log-linear in functional form in our econometric estimation.

Intuitively, equation (8) states that the rate of growth of UK productivity depends on two terms. The first term (γ_i) captures the direct effect of various economic variables (such as for example, domestic R&D intensity and human capital) on the growth rate and their effect on productivity levels through ω.[18] The second term implies that, other things being equal, a sector will enjoy a higher rate of productivity growth the greater the gap between UK productivity and that in the frontier economy (defined here as the US). The function (λ) determines the rate at which productivity in the UK converges to the US, and is allowed to be a function of both the level of openness in each sector and of a variety of other economic factors that may accelerate the rate of convergence, such as the level of human capital. The theoretical model described above and the results are explained in more detail in Cameron, Proudman and Redding (1998b).

One of the most important features of the model of technology transfer described in equation (8) is the level of productivity relative to the technological leader. We therefore begin with a brief discussion of the measurement of and behaviour of levels of productivity in UK manufacturing. The US is assumed to be the technological leader in each industry over the period, and UK productivity relative to the US is measured using an index which, under the assumption of perfect competition, approximates any constant returns to scale production technology.

However, in order to measure relative productivity one must first convert values of output and physical capital into a common currency. The exchange rate chosen is therefore, in principle at least, central to relative productivity measurement. Conceptually, the appropriate exchange rate is the purchasing power parity exchange rate (PPP), which represents the number of dollars required to buy the same quantity of goods as may be purchased with one pound sterling. However, since relative prices may vary significantly across manufacturing industries, it could be misleading to use a single, economy-wide PPP. The approach taken here is to use industry specific purchasing power parities, based upon relative prices in each industry. Specifically, we use the industry-specific PPPs presented in Van Ark (1992), derived from unit value price ratios for a variety of individual products within each manufacturing sector.[19]

An alternative approach (see, for example, Jorgenson and Kuroda, 1990) is to derive industry PPPs from the data on consumer expenditure PPPs for 153 products reported in the United Nations' International Comparisons Project (see Kravis, Heston and Summers, 1988). Each approach has its advantages and disadvantages. Although we favour the unit value-based approach (given the difficulties in correcting for indirect taxes and distribution margins using expenditure PPPs and the absence of information on intermediate inputs), we replicated our estimates of relative productivity using four additional sets of disaggregated PPPs: the results are not substantively sensitive to the choice of PPP. More detail is provided by Cameron, Proudman and Redding (1998b).

Having converted outputs and inputs into a common currency, growth accounting techniques may be used to approximate relative *levels* of TFP in sector j at any given point in time. Here we follow Denny and Fuss (1983) and Bernard and Jones (1996b) in employing an interspatial Divisia index. Under the assumptions of perfect competition and constant returns to scale, relative productivity levels may be approximated by,

$$\ln \tilde{A}(t) \equiv \ln\left(\frac{A_{Bj}(t)}{A_{Fj}(t)}\right) = \ln\left(\frac{Y_{Bj}(t)}{Y_{Fj}(t)}\right) - \frac{1}{2}(\alpha_{Bj}(t) + \alpha_{Fj}(t))\ln\left(\frac{L_{Bj}(t)}{L_{Fj}(t)}\right)$$
$$-\left(1 - \frac{1}{2}(\alpha_{Bj}(t) + \alpha_{Fj}(t))\right)\ln\left(\frac{K_{Bj}(t)}{K_{Fj}(t)}\right) \qquad (9)$$

where, $\alpha_{ij}(t)$ is the share of labour in total income in sector j of economy i at time t.

At this stage, in order to have comparable data, we match the twenty sectors of the UK manufacturing data set discussed in section 2 with the fourteen sectors of the US manufacturing data set. This necessarily leads to a less detailed analysis, but retains the principal differences between the sectors. One important change is that computing is now classified as part of the machinery sector in what follows.

The resulting relative TFP levels are summarised in table 6.6, which presents levels of relative TFP in 1970 and 1990, and average (logarithmic) rates of growth of relative TFP over the periods 1970 to 1990, 1970 to 1979, and 1980 to 1989. Two points emerge from inspection of the table. First, there are substantial variations in relative productivity *levels* across industries: in 1970, paper and printing had the lowest level of relative TFP (40.4 per cent); less than half that in the industry with the highest level (82.0 per cent in machinery). Second, there were substantial changes in the rankings of industries during the period. Between 1970 and 1990, relative TFP in transport rose from 46.7 per cent to 73.4 per cent (an annual average growth rate of 2.25 per cent); while relative TFP in food and drink fell from 72.1 per cent to 57.3 per cent of the US level (an annual average rate of growth of –1.15 per cent).

Figures 6.1–6.4 plot the evolution of TFP in the UK relative to the US. Over the period from 1970 to 1992, TFP in aggregate UK manufacturing rose from around 52 per cent of the US level to roughly 61 per cent, implying a closing of roughly 15 per cent of the productivity gap with the US over the twenty-two year period.

From figures 6.1–6.4 and table 6.6, it is clear that the rate at which the productivity gap was closed varied over the sample period. In total manufacturing, at the end of the 1973–9 peak-to-peak business cycle, there was very little change in UK relative productivity from its 1973 level. In contrast, over the 1979–89 business cycle, the level of UK relative productivity rose from about 53 per cent of the US level to about 58 per cent, an average annual rate of growth of just under 1 per cent. This rise in the rate of

Table 6.6 *Levels and rates of growth of relative TFP in UK and US.*

Industry	TFP_{70}	TFP_{90}	$\overline{TFP}_{70-90}$	$\overline{TFP}_{70-79}$	$\overline{TFP}_{80-89}$
Food, drink, tobacco	72.1	57.3	−1.15	−1.04	−1.37
Textiles and clothing	51.7	58.0	0.57	0.40	1.24
Timber and furniture	50.5	53.5	0.28	0.63	1.20
Paper and printing	40.4	48.9	0.95	−0.56	2.15
Minerals	76.5	76.3	−0.02	−0.81	1.26
Chemicals	49.5	64.0	1.28	1.91	1.28
Rubber and plastics	74.8	90.8	0.97	0.08	2.03
Primary metals	51.5	71.8	1.66	−5.18	10.91
Metal products	41.7	61.1	1.91	1.89	2.77
Machinery	82.0	77.9	−0.32	−0.43	0.73
Electrical engineering	60.6	57.4	−0.27	−1.10	0.28
Transport	46.7	73.4	2.25	−0.22	4.69
Instruments	64.3	76.2	0.85	1.27	−0.96
Other manufacturing	41.2	49.1	0.88	2.41	0.47
Total manufacturing	52.3	61.2	0.79	0.18	1.27

Note: All figures expressed as percentages, growth rates are logarithmic growth rates.

growth of relative productivity between the two peak-to-peak business cycles is consistent with the earlier documentation of the rise in the UK's domestic rate of TFP growth.

Figures 6.2–6.4 compare the evolution of relative TFP for each of the disaggregated manufacturing sectors, grouping sectors by levels of initial UK productivity relative to the US. Clearly, the time series behaviour of different sectors exhibits some variation, but one feature stands out fairly clearly from the figures. The rate at which relative productivity converges to US levels appears to be higher in sectors with low initial levels of relative productivity.

This visual evidence is confirmed if one runs a cross-section regression of average annual rates of growth of relative TFP between 1970 and 1992 against 1970 levels of relative TFP (a test for so-called absolute β-convergence). The estimated coefficient on the initial level of TFP is indeed negative and statistically significant: the rate of productivity catch-up across sectors was inversely related to the initial level of relative productivity. That is, in terms of the convergence literature, relative productivity exhibited absolute β-convergence.[20] The estimated coefficients are shown in table 6.7.

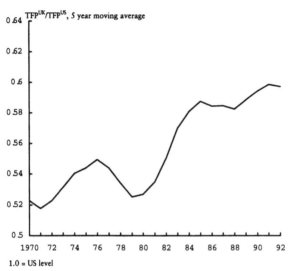

Figure 6.1 *The evolution of TFP in aggregate UK manufacturing relative to the United States*

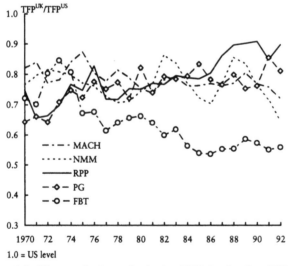

Figure 6.2 *The evolution of relative TFP in the five UK manufacturing sectors with the highest initial level of TFP*

In our estimation, we wish to allow the rate of growth of TFP in a sector to be a function of the levels of a number of variables which, in addition to openness, we believe may be important for explaining the

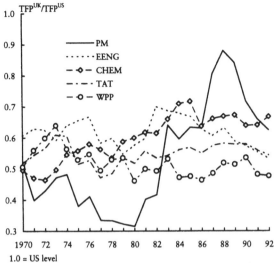

Figure 6.3 *The evolution of relative TFP in the five UK manufacturing sectors with intermediate initial levels of TFP*

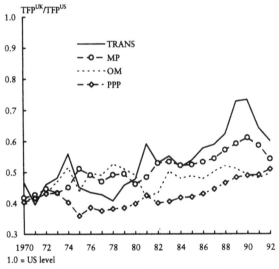

Figure 6.4 *The evolution of relative TFP in the four UK manufacturing sectors with the lowest initial level of TFP*

growth process (and represented in equation (8) by the term (γ_i)). These variables include the intensity of commercially funded research and development and levels of human capital. We also include the degree of trade

Table 6.7 *Cross-section regression of average relative TFP growth (1970–92) on 1970 values of relative TFP*

Dependent variable	Average annual relative TFP growth
Constant	−0.9812**
	(0.4894)
Initial relative TFP	−0.0232**
	(0.0078)
R^2	0.4253
Standard error of regression	0.0069

Notes: ** indicates significance at the 95% level; * indicates significance at the 90 per cent level.

unionisation which some models predict may affect the rate of innovation or technological adoption. We also allow for changes in capacity utilisation, and for the ratio of input to output prices which may distort the estimation of TFP. Not only does this permit a more accurate specification of the growth process, but it also allows us to explore the robustness of the association between productivity growth and openness to the inclusion of a variety of other economic variables.

The main focus of this chapter is whether the rate of productivity convergence λ_j is a function of international openness. We capture this econometrically by including more than one productivity gap term. One is simply the size of the productivity gap: the level of US TFP relative to the UK. The others are the size of the productivity gap multiplied by the level of those variables – including international openness – that may enhance the rate of technology transfer.[21] A positive coefficient on the first term implies that sectors with high initial levels of US TFP relative to the UK grow more rapidly; while a positive coefficient on the openness interaction term implies that more open sectors converge more rapidly towards the technological frontier for a given size of the technological gap.

Our data set includes both the time series and cross-section dimensions, and a total of about 300 observations. To estimate the model on this data set, Fixed Effects Panel estimation was used, which effectively pools observations across sectors and over time but allows for differences between sectors by estimating separate constant terms for each. Within this framework, the model was estimated using least squares.[22] The precise specification of the model[23] was reached by a General to Specific method, starting with a general model and deleting insignificant variables to reach a parsimonious model. The most notable variable we were able to exclude was the degree of trade unionisation.

Reflecting the variety of measures of openness corresponding to the flow of goods, ideas and capital, the system was estimated *separately* for each measure of openness in an otherwise identical regression. The incorporation of the time series dimension as well as the cross section dimension of the data set and the more formal econometric specification of the relationship between openness and growth permits us to distinguish between the effects of different measures of openness in a way we were unable to in section 3.

In table 6.8, we report the regression results for the export and the import ratios, with and without a lagged dependent variable. Identical regressions were run – but are not reported here – measuring openness as the inward and outward foreign direct investment ratios and as the trade-weighted R&D stock ratio. We found that the coefficient on the openness interaction term ('ln(openness interact(−1)') is correctly signed and significant at the 5 per cent level when estimated using the export ratio and the import ratio. The openness interaction term is also correctly signed when openness is measured using the trade-weighted R&D ratio. These results suggest that trade in goods and the flow of ideas are channels through which technology transfer occurs. However, the coefficient on the openness interaction term is insignificant when estimated using either the inward or the outward foreign direct investment ratio. This implies that although FDI is positively correlated with TFP growth across sectors, this correlation does not persist once one controls for the size of the technology gap in a sector and includes a number of other determinants of economic growth.

Turning to the other variables in the model, the coefficient on the productivity gap term ('ln(gap TFP(−1)') is significant and correctly signed as positive, consistent with the hypothesis that the model of technological transfer is an appropriate framework within which to analyse productivity growth. Domestic R&D intensity (Business Enterprise R&D spending divided by value-added) is also significant at the 5 per cent level and the measure of human capital (given by the ratio of workers with high and medium qualifications to total workers) is positive and significant at the 5 per cent level when interacted with the productivity gap (hence implying that higher levels of human capital accelerate the speed of technology transfer). We also find that the change in capacity utilisation is a significant influence on TFP growth, probably reflecting cyclical factors. The magnitude of the coefficients reported in table 6.8 are pretty stable using alternative measures of openness.

The presence of the lagged dependent variable in regressions 1 and 2 results in the familiar Nickell bias to the parameters of interest as discussed

Table 6.8 *Fixed effects panel data least squares estimation: dependent variable: log UK TFP growth*
Sample period, 1970–92. Total panel observations 294.

Dependent variable:		EXPORT/OUTPUT	IMPORT/OUTPUT	EXPORT/OUTPUT	IMPORT/OUTPUT
Regression no:		1	2	3	4
ln (openness /interaction(−1))	(α_1)	0.0886** (0.0375)	0.0426** (0.0221)	0.0780** (0.0347)	0.0394** (0.0198)
ln (gap(−1))	(α_2)	0.2144** (0.0399)	0.1939** (0.0376)	0.2178** (0.0376)	0.2025** (0.0357)
ln (R&D intensity (−1))	(α_3)	0.0323** (0.0149)	0.0320** (0.0149)	0.0350** (0.0142)	0.0352** (0.0142)
ln (human capital interaction (−1))	(α_4)	0.0882** (0.0402)	0.0750* (0.0430)	0.0899** (0.0389)	0.0788* (0.0413)
Δln (capacity utilisation (−1))	(α_5)	−0.0910** (0.0156)	−0.0917** (0.0157)	−0.0904** (0.0139)	−0.0908** (0.0139)
ln (input/output prices (−1))	(α_6)	−0.0896** (0.0368)	−0.0955** (0.0374)	−0.0901** (0.0349)	−0.0942** (0.0356)
Δln UKTFP(−1)	(α_7)	−0.0672 (0.0575)	−0.0730 (0.0576)		
Fixed effects:	($\alpha_{i,0}$)				
Food, drink & tobacco		0.0956	0.0664	0.1024	0.0778
Textiles & clothing		0.0838	0.0761	0.0935	0.0877
Timber & furniture		0.1720	0.0810	0.1774	0.0970
Paper & printing		0.1031	0.0399	0.1051	0.0497
Minerals		0.1127	0.1074	0.1236	0.1212
Chemicals		−0.0233	−0.0045	−0.0204	−0.0041
Rubber & plastics		0.1381	0.1337	0.1441	0.1423
Primary metals		0.0364	0.0330	0.0347	0.0316
Metal products		0.0774	0.0818	0.0839	0.0902
Machinery		0.0517	0.0637	0.0609	0.0721
Electrical engineering		−0.0232	−0.0161	−0.0197	−0.0139
Transport		−0.0578	−0.0327	−0.0560	−0.0334
Instruments		0.0324	0.0504	0.0378	0.0536
Other manufacturing		−0.1074	−0.0517	−0.0908	−0.0443
R^2		0.2588	0.2538	0.2619	0.2591
Adjusted R^2		0.2045	0.1991	0.2132	0.2102
SE of regression		0.0634	0.0636	0.0630	0.0631
Log likelihood		692.1460	695.3335	724.3306	727.8821
Durbin–Watson statistic		1.9275	1.9124	2.053	2.0559
Mean dependent variable		0.0138	0.0138	0.0131	0.0131
SD dependent variable		0.0711	0.0711	0.0710	0.0710
Sum squared residual		1.0958	1.1059	1.1437	1.1481
F-statistic		15.8867	15.4763	20.4346	20.1392
Prob (F-statistic)		0.0000	0.0000	0.0000	0.0000

Notes: ** denotes significance at the 95% level; *denotes significance at the 90% level; (−1) denotes the first lag of a variable. Heteroscedasticity Consistent Standard Errors in parentheses.

earlier. Nonetheless, the lagged dependent variable is not significant at conventional critical values, and our preferred specifications (shown for the import/output and export/output measures of openness in regressions 3 and 4) exclude this variable. A comparison of regressions 1 and 3 (or 2 and 4) provides an assessment of the likely size of the bias: it is clear from table 6.8 that the point estimates of the parameters on the openness terms are essentially unchanged.

A further econometric concern is the potential endogeneity of the explanatory variables. In order to mitigate this problem, our preferred model uses lagged values of the variables of interest. As an additional check, regressions 3 and 4 were re-estimated using Two-stage Least Squares. For each of the explanatory variables, with the exception of capacity utilisation, the first and second lag of the variable were used as instruments. For the change in capacity utilisation, two lags of the change in the UK bank rate, the change in capacity utilisation in total manufacturing, and the change in competitiveness (relative producer prices) were used as instruments (each of which should be exogenous for a particular industry). The parameter estimates were essentially unchanged; although, as is to be expected, there is some loss of efficiency.

We carried out one further robustness check. Regressions 3 and 4 were re-estimated, dropping one industry at a time from the sample. This provides a check that the parameter estimates are not driven by a single influential industry. For both the export/output and import/output regressions and for each of the fourteen industries, the new parameter estimates lay within the 95 per cent confidence interval of the estimates in regressions 3 and 4.

The model of technology transfer described in equation (8) and estimated in table 6.8 implicitly incorporates a long-run steady-state level of output in each sector relative to the US. In the steady-state, the rate of growth of TFP in the UK will equal that in the US. Therefore by setting the growth rate of TFP in the UK – the left-hand side of equation (8) – equal to the estimated long-run rate of growth of TFP in the corresponding US sector, we can back out an expression for the steady-state – or long-run – level of productivity in the UK relative to the US.[24] Rearranging equation (8) and denoting steady-state values with a star yields the following expression,

$$\ln(\frac{A}{A^{US}})^* = \frac{\Delta \ln(A^{US^*}) - \gamma_i (\text{R\&D etc})}{\lambda \,(\text{openness, human capital})} \qquad (10)$$

It follows that, in the long run, the level of productivity in the UK relative to the US tends to a constant which is a function of the rate of catch-up (λ, which is a function of openness and human capital), and the rate of domestic innovation (γ_i) which is a function of the levels of the other significant explanatory variables (domestic R&D intensity, and the input/output price ratio). The openness interaction implies that openness accelerates the rate of productivity growth in the transition to the steady-state (through the rate of convergence) *and* increases the long-run steady-state level of relative productivity. The same is true of human capital. In the long run, the UK's rate of productivity growth equals that of the US.[25]

We can therefore make use of the estimated coefficients from table 6.8 to make inferences about changes in the implicit steady-state level of productivity over the sample period. In the following, we derive the steady-state using coefficients taken from the model estimated using the import ratio. We estimate the implicit steady-states in both 1970 and 1990 and present the results in table 6.9. Taking the average of the fourteen sectors, the steady-state level of productivity in UK manufacturing rose from roughly 58 per cent of US levels in 1970 to some 69 per cent in 1990.

This average conceals considerable variation across sectors. What is clear from the estimates, however, is the considerable increase in the steady-state level of relative TFP across almost all sectors over the period. In only two sectors (timber, and furniture and professional goods) did the steady-state level fall over the period.

Which factors contributed to the observed rise in the steady-state over the period from 1970 to 1990? The contribution of changes in each of the explanatory variables to the increase in the steady-state may be approximated by simulating the steady-state using 1990 values of each explanatory variable in turn, holding all others constant at their 1970 level. Some 51 per cent of the rise in the steady-state level of productivity over the period was attributable to the increase in openness (as measured by the import ratio). Some 55 per cent of the increase in the average steady-state was attributable to the increase in human capital. A fall of about 17 per cent was attributable to changes in R&D intensity in UK manufacturing sectors, while the fall in the ratio of input to output prices made a small positive contribution to the average steady-state.

5 Conclusion

This chapter has examined the extent to which variations in rates of economic growth over time and across sectors in the UK are related to

Table 6.9 *Actual and steady-state levels of UK TFP relative to the US at the start (1970) and end (1990) of the sample period. Steady-state levels derived from coefficients estimated using export/output ratios.*

Sector:	Relative TFP in 1970		Relative TFP in 1990	
	Actual	Steady-state	Actual	Steady-state
Food & drink	0.721*	0.553	0.573	0.674
Textiles	0.517*	0.578	0.580	0.583
Timber & furniture	0.505	0.556	0.535	0.576
Paper & printing	0.404	0.454	0.489	0.530
Minerals	0.765*	0.717	0.763	0.826
Chemicals	0.495	0.573	0.640	0.785
Rubber & plastics	0.748	0.819	0.908	0.921
Primary metals	0.515	0.538	0.718*	0.669
Metal products	0.417	0.517	0.611	0.705
Machinery	0.820*	0.724	0.769	0.860
Electrical engineering	0.606*	0.517	0.574	0.701
Transport	0.467*	0.463	0.734*	0.663
Instruments	0.643	0.814	0.762	0.784
Other manufacturing	0.412	0.434	0.491	0.572

Note: * denotes a sector in which actual relative TFP exceeds estimated steady-state relative TFP

differences in the degree of international openness. Openness has three main dimensions: international trade in goods and services, international movements in factors of production and the international spillover of ideas.

An important feature of the chapter has therefore been to compile and estimate accurate measures of productivity and openness at a disaggregated level. In particular, two data issues stand out. First, there are problems associated with the potential endogeneity of measures of openness. We have used a variety of econometric procedures (including lagged openness variables and instrumental variables) to deal with this, and shown that our results are robust to the use of alternative techniques. Second, we have taken a careful approach to the construction of the data, paying particular attention to problems with deflators and purchasing power parities.

The recent theoretical literature on endogenous growth provides three main proximate mechanisms through which openness may affect growth. International openness may affect either the domestic rate of innovation, or the amount of technology that can be transferred, or the rate of adoption of technologies from more advanced countries. It is not possible to identify the former two influences separately in our model. We therefore test whether openness affects (a) the rate of domestic innovation or the fraction of transferrable knowledge, versus (b) the rate of technology transfer. The main empirical results provide a body of evidence not only to support the claim that openness and productivity growth are closely associated, but also that is consistent with the hypothesis that openness increases the rate of adoption of technology developed in more advanced economies.

More precisely, the main conclusions that have emerged during the course of our research are as follows:

1. Labour productivity in the UK *is* correlated with a number of measures of openness. This correlation is explained by an association between openness and Total Factor Productivity but not capital accumulation. This is consistent with the hypothesis that openness affects growth through the rate of technological change rather than through investment. Indeed, within UK manufacturing, open sectors have recorded markedly higher TFP growth than closed sectors.
2. Much of the growth in aggregate UK manufacturing productivity is due to growth *within* sectors, rather than switches in factor resources between sectors. This suggests that the impact of openness on growth works largely through its effects on the incentives to innovate or adopt technology relative to engaging in current production within sectors, rather than through the reallocation of resources across sectors resulting from changes in comparative advantage.
3. The rate of productivity growth in UK manufacturing *is* associated with differences in the initial level of total factor productivity between the UK and the US. Between 1970 and 1992, some 15 per cent of the initial gap in levels of TFP between the two countries was closed. The rate at which the gap was closed was fastest in sectors with the lowest initial levels of relative TFP. This is consistent with the hypothesis that the adoption of technology is an important determinant of productivity growth.
4. The rate at which a manufacturing sector's productivity converges to the US level depends upon the level of international openness – as measured by the flow of goods and the flow of ideas, but not the flow of capital. That is, trade in goods and the spillover of ideas accelerate the rate at which UK productivity converges to US levels. This finding remains true when we allow for changes in capacity utilisation, the degree of unionisation, the intensity of domestic R&D and the level of human capital.
5. In addition, human capital (as measured by the ratio of workers with high and medium qualifications to total workers) is found to positively

affect the rate of technology transfer. The intensity of domestic R&D (as measured by the ratio of business enterprise R&D to value-added) is found to affect either the rate of domestic innovation or the quantity of knowledge that can be transferred.

Over the sample period, substantial convergence to US levels of productivity occurred in UK manufacturing sectors. In aggregate manufacturing, some 15 per cent of the original productivity gap was closed over the twenty-two year period. Much of this occurred during the 1980s. The convergence in actual values of relative productivity was associated with a rise in the estimated steady-state level of UK relative productivity from about 58 per cent of the US level in 1970 to 69 per cent in 1990. The rise in international openness over the sample period was estimated to account for roughly one half of this increase.

Notes

1. The views expressed are those of the authors and not necessarily those of the Bank of England. This paper was prepared for the NIESR Conference on Productivity and Competitiveness, London, 5–6 February 1998. Cameron's research was funded by the ESRC (grant no. R000234954) and Proudman and Redding's by the Bank of England. We would like to thank Bill Allen, Charlie Bean, Steve Bond, Willem Buiter, Maxwell Fry, Mary Gregory, Zvi Griliches, Neal Hatch, Nigel Jenkinson, Mervyn King, John Muellbauer, Richard Pierse, Chris Pissarides, Danny Quah, Peter Sinclair, Andrew Scott, Jonathan Temple, Peter Westaway, Tony Venables, Tony Yates, and participants at seminars at the London School of Economics, and at a conference on Openness and Growth at the Bank of England for helpful comments. The proceedings of this conference were published by the Bank of England in July 1998. The usual disclaimers apply.
2. OECD Intersectoral Database, based on data provided by the ONS.
3. ONS data.
4. Examples of models of this form include Grossman and Helpman (1991, chapters 11 and 12), Eaton and Kortum (1995) and Parente and Prescott (1994).
5. For a more detailed review of the theoretical literature, see Redding (1997).
6. For an empirical analysis of the dynamics of patterns of international specialisation across UK manufacturing sectors, see Proudman and Redding (1997).
7. See for example, Grossman and Helpman (1991), Aghion, Dewatripont and Rey (1996) and Aghion, Harris and Vickers (1996).
8. More detailed results are contained in Cameron, Proudman and Redding (1997).
9. Since value-added is essentially gross output minus intermediates and the price indices for these components may behave differently, one should deflate gross out-

put and intermediates separately. However, since we are concerned about the quality of the intermediate price indices, we follow Cameron (1996) in using single-deflated value-added.
10 There are some slight differences between these ONS data and the those quoted in the first paragraph of the chapter which come from the OECD Intersectoral Database. For comparisons at the sector level, the ONS are more reliable.
11 See Cameron, Proudman and Redding (1997) p. 28–31.
12 Note that this imposes a more restrictive form of the production function than elsewhere in the chapter.
13 For a theoretical model in which R&D expenditures are an important determinant of growth see Aghion and Howitt (1992). Benhabib and Spiegel (1994) emphasise human capital, while Ulph and Ulph (1994) consider the role of unionisation.
14 There are several forms of discriminant analysis, see Mardia, Bibby and Kent (1979). In this paper we use Fisher's Linear Discriminant Function, which maximises the ratio of the sum of squares of sub-group means to the sum of squares of observations around their sub-group means by taking linear combinations of the openness variables, see Proudman, Redding, and Bianchi (1997).
15 We constructed trade-weighted R&D stocks following Coe and Helpman (1995).
16 Assuming the two samples are drawn from two normally distributed populations with the same variance, we can reject the null hypothesis that the TFP growth rates are the same in each population at the 10 per cent level.
17 The presence of ω_{B_j} (not necessarily equal to 1) generalises the specification in Bernard and Jones (1996a). If $\omega_{B_j} = 1$, then (as will be seen below) all the steady-state gap between UK and US TFP must be explained in terms of differences in sector-specific rates of innovation (γ_{ij}) and the size of the parameter λ_j. Expressed another way, the model implies that UK TFP will in the long-run equal that of the US, in the absence of continuing innovation in the two economies. It is not clear that this is true, since there are differences in levels of human capital or impediments to the flow of ideas that might prevent the UK from attaining US levels of TFP. Therefore we allow ω_{B_j} to differ from 1. Note, however, our theoretical or empirical results are not sensitive to this generalisation of the basic model.
18 Unless identifying restrictions are imposed, it is not possible to distinguish an effect of openness on domestic rates of innovation (γ) separately from an effect on the fraction of transferrable knowledge (ω). Hence we test whether openness affects (a) either the rate of innovation (γ) or the fraction of transferrable knowledge (w), versus (b) the rate of technology transfer (λ).
19 A unit value ratio is simply the ratio of producers' sales values to the corresponding quantities.
20 Note that the finding of absolute β-convergence does not necessarily imply a declining cross-section dispersion of relative productivity. See Quah (1993, 1996) for a discussion of Galton's fallacy.
21 Formally, the terms are $\ln(\frac{A^{US}_{t-1}}{A_{t-1}})$ and $\ln(\text{openness}).\ln(\frac{A^{US}_{t-1}}{A_{t-1}})$.

22 Least squares can potentially generate biased estimates within a Fixed Effects Panel, due to the well-known Nickell bias when there is a lagged dependent variable, see Nickell (1981). The relatively large time dimension of the panel means that the Nickell bias is unlikely to be large. As discussed below, all our regressions were also estimated using an instrumental variables estimator to allow for potential endogeneity. The instrumental variables estimates were little different from their OLS counterparts.

23 The estimated equation is:

$$\dot{A}_{i,t} = \alpha_{i,0} + \alpha_1 \ln(\Omega_{i,t-1})\ln(A^{US}_{i,t-1}) + \alpha_2 \ln(A^{US}_{i,t-1} / A_{i,t-1})$$

$$+\alpha_3 \ln(R_{i,t}) + \alpha_4 \ln(H_{i,t})\ln(A^{US}_{i,t-1}) / A_{i,t-1})$$

$$+\alpha_5 \Delta \ln(C_{i,t}) + \alpha_6 \ln(IO_{i,t}) + \alpha_7 \ln(A_{i,t} / A_{i,t-1}) + \xi_{i,t}$$

Differences between US and UK industrial classifications mean that we can only disaggregate relative TFP into only fourteen sectors rather than the nineteen discussed in section 2.

24 We proxy the long-run rate of growth of TFP in the US in each sector by the sample average annual growth rate of TFP.

25 We do not attempt in this model to determine the long-run world growth rate. However, it is consistent both with the theoretical and the empirical framework outlined above for the long-run joint growth rate to be affected by changes in the degree of openness in the international economy.

References

Aghion, P., Dewatripont, M. and Rey, P. (1996), 'Competition, financial discipline and growth,' University College London, mimeo.

Aghion, P., Harris, C. and Vickers, J. (1996), 'Competition and growth with step-by-step technological progress,' University College London, mimeo.

Aghion, P. and Howitt, P. (1992), 'A model of growth through creative destruction', *Econometrica*, 60, pp. 323–51.

Benhabib, J. and Spiegel, M. (1994), 'The role of human capital in economic development: evidence from aggregate cross-country data,' *Journal of Monetary Economics*, 34, pp. 143–73.

Bernard, A. and Jones, C. (1996a), 'Productivity across industries and countries: time series theory and evidence', *Review of Economics and Statistics*, 78, 1, pp. 135–46.

(1996b), 'Productivity and convergence across US states and regions', *Empirical Economics*, 21, 1, pp. 113–35.

Cameron, G. (1996), 'Innovation and economic growth', DPhil Thesis, University

of Oxford.
Cameron, G., Proudman, J. and Redding, S. (1997), 'Deconstructing growth in UK manufacturing', Bank of England Working Paper No. 73.
— (1998a), 'Openness and its association with productivity growth in UK manufacturing', chapter 5 in Proudman, J. and Redding, S. (eds), *Openness and Growth*, London, Bank of England, pp. 173–211.
— (1998b), 'Productivity convergence and international openness', chapter 6 in Proudman, J. and Redding, S. (eds), *Openness and Growth*, London, Bank of England, pp. 221-260.
Coe, D. and Helpman, E. (1995), 'International R&D spillovers', *European Economic Review*, pp. 859–87.
Denny, M. and Fuss, M. (1983), 'A general approach to intertemporal and interspatial productivity comparisons', *Journal of Econometrics*, 23, 3, pp. 315–50.
Eaton, J. and Kortum, S. (1995), 'International patenting and technology diffusion', NBER Working Paper No. 5207.
Grossman, G. and Helpman, E. (1991), *Innovation and Growth in the Global Economy*, Cambridge, Mass., MIT Press.
Jorgenson, D. and Kuroda, M. (1990), 'Productivity and international competitiveness in Japan and the United States', in Hulten, C. (ed.), *Productivity Growth in Japan and the United States*, Cambridge, Mass., NBER.
Kravis, I., Heston, A. and Summers, R. (1988), *International Comparisons of Real Product and Purchasing Power*, Baltimore, Johns Hopkins University Press.
Mardia, K., Bibby, J. and Kent, J. (1979), *Multivariate Analysis*, London, Academic Press.
Nickell, S. (1981), 'Biases in dynamic models with fixed effects', *Econometrica*, 49, pp. 1417–26.
Parente, S. and Prescott, E. (1994), 'Barriers to technology adoption and development', *Journal of Political Economy*, 102, 2, pp. 298–321.
Proudman, J. and Redding, S. (1997), 'Persistence and mobility in international trade', Bank of England Working Paper No. 64.
Proudman, J., Redding, S. and Bianchi, M. (1997), 'Is international openness associated with faster economic growth?', Bank of England Working Paper No. 63.
Quah, D. (1993), 'Galton's fallacy and the convergence hypothesis', *Scandinavian Journal of Economics*, 95, pp. 427–43.
— (1996), 'Empirics for economic growth and convergence', *European Economic Review*, 40, pp. 1353–76.
Redding, S. (1997), 'Openness and growth: theoretical links and empirical estimation', Bank of England, mimeo.
Ulph, A. and Ulph, D. (1994), 'Labour markets and innovation: ex-post bargaining', *European Economic Review*, 38, pp. 195–210.
Van Ark, B. (1992), 'Comparative productivity in British and American manufacturing', *National Institute Economic Review*, 142, pp. 63–74.

7 Innovation and market value

BRONWYN H. HALL[1]

'Possession of knowledge is worth a thousand pieces of gold.'
Chinese fortune cookie.

Introduction

Private firms and governments share an interest in evaluating the economic returns to their innovative activities. The most common quantitative approach to this measurement problem is to relate total factor productivity or profit growth to measures of innovation (see Mairesse and Mohnen, 1996, for a recent survey of results using this methodology). However, there are a variety of reasons why this approach may be incomplete or difficult to implement in some cases. First, the occasionally long and uncertain lags between spending on innovation and the impact of that innovation on the 'bottom line' mean that in some cases, such as those involving very basic research, the data will not cover a long enough time period to enable precise measurement of the total effect. Second, these same lags mean that one may have to wait for a certain amount of time to see the effects in productivity, making the exercise of limited value for planning purposes. Third, measuring the returns to R&D or other activities using firm or industry level data on profits or output requires careful attention to the measurement and timing of other inputs that may not be possible using available data.[2]

For these reasons, some researchers have turned to another method of evaluating the *private returns* to innovative activity, relating the valuation placed by the financial markets on a firm's assets to its Research and Development expenditure, patenting activities, and other measures of innovation. This method is intrinsically limited in scope, because it can be used only for private firms and only where these firms are traded on a well-functioning financial market (such as in the United States and the United Kingdom, where most of the work has been done to date). Nevertheless, using financial market valuation avoids the problems of timing of costs and revenues highlighted by Fisher and McGowan (1983), and is capable of forward-looking evaluation, which the traditional productivity method does not do well. This method is also potentially useful for calibrating various innovation measures, in the sense that one can measure their economic impact using the widely-available United States firm data, possibly enabling one to validate these measure for use elsewhere as proxies for innovation value.

Interest in valuing innovation assets stems from several distinct sources, and as a result there has been more than one strand of literature: first, firms and their accountants have been anxious to develop methods to value intangible assets of the innovative kind, both to help guide decision-making, and sometimes for the purposes of transfer pricing or even the settlement of legal cases. This has led to consideration of the problem in the financial accounting literature (see, for example, Chauvin and Hirschey, 1993, Lev and Sougiannis, 1996). Second, financial economists and investors often try to construct measures of the 'fundamental value' of publicly traded firms as a guide to investment; a concern with valuing the intangible assets created by R&D and other innovative activities is naturally a part of this endeavour. Finally, policy makers and economists wish to quantify the private returns to innovative activity in order to increase understanding of its contribution to growth and as a guide for strategies to close the gap between private and social returns. A by-product of this goal is the desire to calibrate measures such as patent counts or innovation counts using market-based measures like firm value (see Griliches, Pakes and Hall, 1987, for an earlier survey of this work).

Why might the market value of firms be a useful measure of the private returns to innovation? In a market economy, the private economic 'value' of a good is usually the price at which it trades in the marketplace. In the case of a knowledge asset, this price should embody all the tastes that consumers have for any particular innovation or the knowledge of how to make that innovation. That is, if we want to measure the returns or profit available from an intangible asset, the ideal would be to observe a market on which the asset trades and to measure the price at which the trade takes place. In the case of the output of the R&D performed by private firms, this is quite difficult. We observe consumer demand for particular products, but it is difficult to assign innovative inputs or outputs directly to these products in the absence of very detailed firm data. In any case, the relevant intangible assets that are necessary for producing these products and delivering them to the marketplace usually come bundled in ways that prevent us from separating them and selling them off to determine the appropriate price of a specific individual asset.

For example, the fact that it is not easy to separate the knowledge of how to make a particular chemical entity from the other assets of the pharmaceutical firm that converts this entity into a marketable drug is a problem similar to that of determining the value of factory-installed automobile air-conditioning to consumers by selling it separately from the car in which it was installed. Thus many researchers confronted with this measurement problem use a solution familiar from the automobile demand literature, the hedonic regression method. That is, they try to determine the marginal value of a particular intangible asset by regressing the market price for firms that possess the asset on various characteristics of the firms, including the book value of the intangible asset in which they are interested. Implicitly they assume that financial markets price the bundles of assets that compose a firm (ordinary plant and equipment, inventories, knowledge assets, customer networks, brand names and reputation, and so forth) correctly and that the marginal shadow value (the gross rate of return) of the knowledge asset in the marketplace can be inferred from the regression coefficient estimate.

A few other market-based methods of valuing intangibles are feasible in particular settings: for example, consumer willingness-to-pay for particular innovations was used by Trajtenberg (1989), who studied hospital purchase of CAT scanners. Licensing fees for patents, determined by negotiation between firms or between firms and universities, could be considered market-based measures of expected innovation value (Harhoff *et al.*, 1997). But few measures are available for as wide a range of technologies and industries as firm market value, although, of course, the downside of this measure is its aggregate nature.

The goal of the present chapter is to outline how the market value approach has been used in the past for valuing innovative assets, and to survey some of the results in this literature, as well as presenting some new results for the United States, drawn from two current research projects (Hall and Kim, 1997, Hall, Jaffe and Trajtenberg, 1998). The latter project is particularly novel, as it incorporates citation weights in constructing the patents variable, yielding a somewhat better measure than raw patent counts. There have been hints that this might be a useful measure in the literature earlier, but never using data over such a broad range of industries (see, for example, Trajtenberg, 1990, Shane, 1993, Austin, 1993 and Harhoff *et al.*, 1997).

Measuring the value of knowledge assets

Using firm market value as a measure of innovation returns relies on the fact that publicly traded corporations are bundles of assets (both tangible and intangible) whose values are determined every day in the financial markets. In that sense, they are not different from other goods with heterogeneous characteristics, such as automobiles, personal computers or even breakfast cereal. Since the pioneering work of Waugh (1928), Griliches (1961) and others, hedonic price equations have been widely used to measure the 'prices' of individual characteristics that are bundled into heterogeneous goods. The market value application is not really different from the methodology used in those papers: we are measuring the marginal value of an additional dollar of investment in a given type of corporate asset, using as our data points a set of heterogeneous firms.

The typical model of market value hypothesises that the market value of a firm is a function of the set of assets that it comprises:

$$V(A_1, A_2, A_3, ...) = f(A_1, A_2, A_3, ...) \qquad (1)$$

where f is an unknown function that describes how the assets combine to create value. If the firm invests in the various assets $A_1, A_2, A_3, ...$ according to a value-maximising dynamic programme, and if the stock market is efficient, the function f will be the value function associated with that dynamic programme. In the case with a single asset and constant returns to scale (linear homogeneity) of the profit function, we will obtain the well-known result that the market value V is a multiple of the book value of the asset A, with a multiplier (shadow price) equal to Tobin's q.

Making the comparison to the ordinary hedonic price literature highlights several problems of interpretation or difficulties with this approach:

1. As is well known, the shadow price or hedonic price measures neither supply nor demand of the particular asset; it is a measure of the equilibrium between the two at a point in time. Because it is very far from a structural parameter, there is no reason for it to be stable over time, for example. For the purpose of evaluating expected returns to the investments that have been made, the fact that we are simply measuring the market price of these investments is not a problem (in fact, it is of interest), but it would not be appropriate to treat this market price as an invariant.
2. The functional form of equation (1) is not known, nor is it easy to compute one in closed form if one assumes a realistic profit-

maximising algorithm for the firm. In general, we will fall back on fairly ad hoc functions, such as linear or Cobb–Douglas (linear in the logs).
3. Unlike automobiles, computers, or breakfast cereal, it is sometimes fairly easy to unbundle the corporate assets and trade them separately, which means that we will need an assumption of market efficiency to use a hedonic equation to measure the value of the assets from data on firms. That is, we need to assume that at any point in time, value-increasing unbundling will already have taken place.

Given the difficulty of deriving the value function from an explicit dynamic program or maximisation model (see Wildasin, 1984, and Hayashi and Inoue, 1991, for solutions to some simple models when there is more than one type of asset), empirical workers have fallen back on several simple solutions guided by the theory and basic econometric considerations, in much the same way that hedonic price equations have been constructed for other durable goods. A central question is whether the assets in a firm can be treated as additively separable (which implies that the firm is equal to the sum of its parts, or alternatively, that it would be possible to unbundle the assets and sell them separately for the same price they fetch when embedded in the firm) or whether a more complex multiplicative functional form must be used. In spite of the obvious unattractiveness of the additively separable function, it has been widely used because of its simplicity, following the initial work of Griliches (1981).[3]

Thus the following two specifications of the value function are predominant in the literature: an additively separable linear specification, as was used by Griliches (1981) and his various co-workers, and then a multiplicative separable specification of the Cobb–Douglas form. These two forms differ in that the additively separable version assumes that the marginal shadow value of the assets is equalised across firms, while the Cobb–Douglas version assumes that the value elasticity is equalised.[4]

The first (linear) model is given by

$$V_{it}(A,K) = q_t (A_{it} + \gamma_t K_{it})^{\sigma_t} \qquad (2)$$

Taking logarithms of both sides, we obtain

$$\log V_{it} = \log q_t + \sigma_t \log A_{it} + \sigma_t \log(1 + \gamma_t K_{it} / A_{it}) \qquad (3)$$

In most of the work reported on here,[5] the last term is approximated by $\gamma_t K_{it} / A_{it}$, in spite of the fact that the approximation can be relatively inaccurate for *K/A* ratios of the magnitude that are now common (above 15 per cent). In this formulation, γ_t measures the shadow value of R&D assets relative to the tangible assets of the firm and $q_t \gamma_t$ measures their absolute value (when σ_t is approximately unity).

The second (log-linear) model has the Cobb–Douglas form:

$$V_{it}(A,K) = q_t A_{it}^{\sigma_t - \alpha_t} K_{it}^{\alpha_t} \qquad (4)$$

In logarithms, this equation is the following:

$$\log V_{it} = \log q_t + \sigma_t \log A_{it} + \alpha_t (\log K_{it} / A_{it}) \qquad (5)$$

In both models, the coefficient of *A* is unity under constant returns to scale or linear homogeneity of the value function. If the assumption of constant returns is true (as it will be approximately in the cross section), it is possible to move the log of ordinary assets to the left hand side of the equation and estimate the model with the logarithm of the conventional Tobin's q as the dependent variable. The intercept of either model can be interpreted as an estimate of the logarithmic average of Tobin's q for the sample of firms during the relevant period. In order to compare the results of the second model to the results of the first, we need to compute the ratio of the marginal shadow value of *K* to that of *A*:

$$\frac{\partial V / \partial K}{\partial V / \partial A} = \frac{\alpha_t V_{it} / K_{it}}{(\sigma_t - \alpha_t) V_{it} / A_{it}} = \frac{\alpha_t A_{it}}{(\sigma_t - \alpha_t) K_{it}} \qquad (6)$$

This measure can be compared to the γ_t estimated by the first model, but to do so we will need to use some kind of average value of *A/K*. The absolute shadow value of R&D capital is equal to $\alpha_t V_{it} / K_{it}$.[6]

In passing, we can note that a variant of equation (2) has been used by some researchers (notably Connolly, Hirsch and Hirschey, 1986, and Connolly and Hirsch, 1988), where constant returns ($\sigma_t = 1$) and market equilibrium ($q_t = 1$) are imposed and A_t is subtracted from the left hand side to give the following:

$$V_{it}(K) - A_{it} = \gamma_t K_{it} \qquad (7)$$

In this case the excess of market value over book value of the assets is regressed on various measures of intangibles. An obvious difficulty with this version is that if Tobin's q differs from unity on average (as it almost always does), $(q_t - 1)A_{it}$ will end up in the disturbance and potentially bias the estimates of γ_t (which is an estimate of either the relative or the absolute shadow value of *K* in this formulation).

Market value and R&D

Table 7.1 presents a summary of some of the earlier work relating the market value of individual firms to innovation indicators such as R&D and patenting. Most of this prior literature has used US data and the linear form of the value equation. With the exception of Hall (1993a, 1993b), Chauvin and Hirschey (1993), and Stoneman and Toivanen (1997), parameter stability has been imposed on the relationship over time. The table shows that researchers using data for United States manufacturing firms generally conclude that R&D spending in the current year is capitalised into the market value at a rate between about 2.5 and 8 (with most estimates centred at 5 to 6) and that the stock of R&D (which is usually constructed from a perpetual inventory formula using a depreciation rate of 15 percent) is valued between 0.5 to 2 times the value of ordinary assets. A notable exception to the latter result is that of Jaffe (1986), who finds a much higher R&D stock coefficient when he controls for the potential spillovers from firms that patent in related technology fields. In general, the addition of industry dummies (at the two-digit level) to the equation does not change the estimates, although including firm dummies or lagged 'q' does lower the R&D coefficient, which suggests that there are some permanent differences in the market value–R&D relationship across firms.

The focus of the particular studies summarised in table 7.1 has varied: for example, Jaffe (1986) was an innovative and careful investigation into the contribution of the R&D of other firms that are in the same technology space as the firm in question to its patenting, profits and market value. He found that the raw contribution to market value was rather weak and slightly negative, but that the contribution was positive and significant when the firm in question had a good-sized R&D programme of its own. Connolly, Hirsch and Hirshey (1986) were concerned with the effects of unions on the incentives to undertake R&D and on the returns obtained from that R&D; they found, as expected, that firms in unionised industries were less likely to perform R&D and that when they did, the rents received were lower.

Cockburn and Griliches (1987) attempted to use measures of appropriability from the Yale survey on innovation to explain variations in the shadow value of R&D in different firms and industries with very limited success, while Megna and Klock (1993) focused on the effects of rivalry in R&D in a specific industry (semiconductors).

Most of the earlier US results reported use outturns for the late 1960s and 1970s. Later work by Hall (1993a) and others (for example, Hirschey, Richardson and Scholz, 1998) revealed both that this was a period of

Table 7.1 *Market value – innovation studies with R&D and patents*

Study	Country (industry)	Years	Functional form	Other variables	R&D coeff.	R&D stock coeff.	Patent or innov. coeff.	Comments
Griliches, 1981	US	1968-74	Linear (Q)	Time & Firm dummies, [log Q(-1)]	1.0-2.0		0.08 to 0.25	units appear to be 100 pats
Ben-Zion, 1984	US	1969-76	Linear (V)	Ind dummies, Investment, Earnings	3.4 (0.5)		0.065 (0.055)	
Jaffe, 1986	US	1973, 79	Linear (Q)	Time & tech dummies, C4, mkt share, Tech pool, interactions		7.9 (3.3)		3SLS even higher
Hirschey, Weygandt, 1985	US	1977	Linear (Q)	Adv, C4, growth, risk	8.3 (1.4)			Compare durable/non-dur.
Connolly, Hirsch, Hirschey, 1986	US	1977	Linear (EV/S)	Growth,risk,age,Mkt share,C4, Adv, Union share, Ind dummies	7.0 (0.8)		4.4 (0.6)	Unexpected patents
Cockburn, Griliches, 1987	US		Linear (Q)	Industry appropriability (Yale survey)				
Connolly, Hirschey, 1988	US	1977	Linear (EV/S)	Growth, risk, C4, Adv	5.6 (0.6)		5.7 (0.5)	Bayesian estimation
Connolly, Hirschey, 1990	US	1977	Linear (MV/S)	Adv, MS, C4, risk, growth, inv	5.7 (0.7)		5.7 (0.5)	Patent surprise
Hall, 1993a	US	1973-91 Gr, time dummies	Linear (V)	Assets, Cash flow, Adv, (0.8)	2.5-3.0 (0.02)	0.48		By year also
Hall, 1993b	US	1972-90	Linear (Q)	Time dummies	2.0-10.0	0.5-2.0		By year; LAD; absolute coeff.

Table 7.1 (continued)

Study	Country (industry)	Years	Functional form	Other variables	R&D coeff.	R&D stock coeff.	Patent or innov. coeff.	Comments
Megna, Klock, 1993	semi-conductors	1977-90	Linear (Q)	Rivals R&D and patents		0.82 (0.2)	0.38 (0.2)	Patent stock
Chauvin, Hirschey, 1993	US	1988-90	Linear (V)	Cash Flow, growth, risk, adv, MS, ind dums	6.47 (0.35)			Compares size, non-mfg/mfg
Blundell, Griffith, van Reenen, 1995	UK	1972-82	Linear (V)	Time dummies, Assets, Mkt share			1.93 (0.93)	Innovation counts
Stoneman, Toivanen, 1997	UK	1989-95	Linear (V)	Assets, Debt, Growth, Mkt share, investment, Cashflow, time dummies, Mills ratio	2.5 (1.5)		insig.	Selection correction; by year
Chauvin, Hirschey, 1997	US	1974-90	Linear (V)	Cash flow, beta, growth-Adv, growth-MS interaction, ind dummies	1.7 (0.5)			can't derive R&D coeff.
Hirschey, Richardson, Scholz, 1998	US	1989-95	Linear (Q)	Earnings, R&D and patents		0.20 (0.06)	3.30 (0.65)	Current impact, sci link, deprec

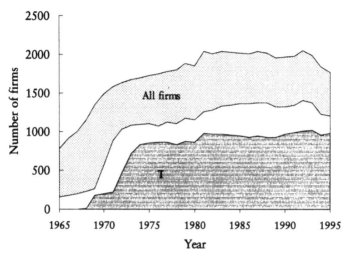

Figure 7.1 *Cleaned sample of US manufacturing firms*

relatively high valuation of R&D in the market and that the shadow value of R&D does not display much stability over time. Hall (1993a, 1993b) found that the relative valuation coefficient had declined rather abruptly during the 1980s in the United States, and that the decline could be partially explained by two factors:

1. an increase in the value of ordinary assets, probably associated with the pervasive restructuring that took place during the period, and
2. a decline in the value of R&D assets that appeared to be concentrated in the electrical machinery, computing, electronics, and scientific instruments sector.

This result was consistent with what we know about the pace of technical change in those industries, and suggested that the private returns to R&D done in those industries were not very long-lived. However, the size of the decline raised questions about the incentives for innovation in US manufacturing in the late 1980s and early 1990s; clearly the behaviour of firms suggested that many firms still viewed R&D as a profitable activity, in spite of the apparent market signal.

Current research by Daehwan Kim and the author updates these results to 1995, and explores their robustness to changes in functional form, improvements in measurement of the other assets variables, and so forth. Figure 7.1 displays the sample of firms being used, which consists of es-

Innovation and market value

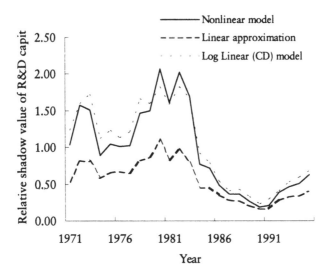

Figure 7.2 *Firms doing R&D – trimmed sample – comparing specifications*

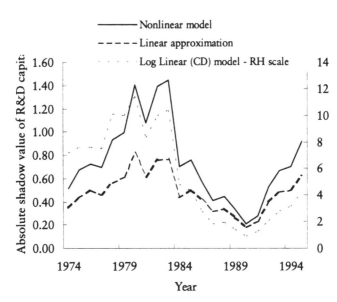

Figure 7.3 *Firms doing R&D – trimmed sample – comparing specifications*

sentially the entire publicly traded US manufacturing sector.[7] Figure 7.2 displays preliminary results obtained using both equations (5) (log linear) and (1) (linear in the stocks, in either a nonlinear or linear-in-variables form of the equation). We have computed the relative marginal product that corresponds to the coefficients from equation (5) using equation (6). The figure shows that the decline in the valuation of R&D assets observed through 1991 has begun to recover in the mid-1990s, although not to the level of the boom years of the early 1980s. Nonnested hypothesis testing suggests that the data prefer either the log linear (Cobb–Douglas) model or the nonlinear model, with the linear approximation to the nonlinear model a poor third. These results should be seen as preliminary, and various other dimensions are being explored.

There appear to be only two studies of the market value of innovative assets that use non-US data, both for the United Kingdom: Blundell, Griffith and van Reenen (1995), who use innovation data rather than R&D, and Stoneman and Toivanen (1997) who use R&D data for 1989 to 1995.[8] Blundell et al. is discussed in the next section along with the review of the results using patents rather than R&D. Because required reporting of R&D has just begun in the United Kingdom, Stoneman and Toivanen are careful to estimate the relationship between market value and R&D using sample selection methods to correct for the probability of reporting R&D. They are also unable to construct a stock of R&D capital owing to the short history of spending for most firms, so they report estimates using a flow measure of R&D instead. They find that the coefficient of R&D spending varies over time for their firms, with a range from zero to 4.3; in most years it is significant and the average value seems to be about 3, which is slightly lower than the corresponding number for the United States in the early nineties (approximately 1 to 2). They also find the very interesting result that when a firm first announces its R&D (i.e., when it begins reporting the figure publicly), the multiplier is somewhat higher, with a value close to 5. This suggest that the timing of the announcement by the firms is perhaps chosen by the firm in order to have maximal impact on its market value.

Market value and patents

Research that uses patents in the market value equation rather than R&D is somewhat more limited, primarily because of the difficulty of constructing firm data sets that contain patent data. Most of the work shown in table 7.1 and described here has been done by Griliches and his co-workers using a database constructed at the NBER which contained data on patents only through 1981. This data set did not include information on the citations related to the patents. The other papers in the table use either a cross section constructed by Connolly and others for 1997 of Fortune 500 companies, or data sets involving UK data that includes innovation counts from a SPRU study rather than direct registration of patents.

When patents are included in a market value equation, they typically do not have as much explanatory power as R&D measures, but they do appear to add information above and beyond that obtained from R&D, as one would expect if they measure the 'success' of an R&D programme. Griliches, Hall and Pakes (1991) show that one reason patents may not exhibit very much correlation with dollar-denominated measures like R&D or market value is that they are an extremely noisy measure of the underlying economic value of the innovations with which they are associated. This is because the distribution of the value of patented innovations is well known to be extremely skewed, implying that a few patents are very valuable, and many are worth almost nothing. Therefore the number of patents held by a firm is a poor proxy for the sum of the value of those patents and we should not expect the correlation to be high. Some small studies exist that suggest that the number of citations received by a patent may be correlated with its economic value (Trajtenberg, 1990, Shane, 1993, Harhoff *et al.*, 1997), so that weighting patents by the number of citations they receive may improve the measure.

There is also good reason to think that the meaning of a patent to a US corporation may have changed somewhat since 1981 (see the results of the Carnegie-Mellon or Yale II survey, which differed in some sectors from those of the Yale I survey of the early 1980s). For both these reasons, Jaffe, Trajtenberg and Hall have embarked on a project to create a new data set of US firms that contains data on all the patents held by the firms, including information on the citations received. This paper reports preliminary results from that study.

Figure 7.4 shows the fraction of firms in the sample in a given year who reported R&D expenditures to the SEC, the fraction who applied for a patent that was ultimately granted, and the fraction who have a nonzero stock of patents that year.[9] The fall in the later years in the number of firms with patent applications is due to the fact that patents may have been applied for but not yet granted. Figure 7.5 shows the ratio of citations made to patents held by firms in our sample to the number of patent grants, all dated by the application year of the patents, and the median citation count per patent at the firm level. Both these numbers also fall off begin-

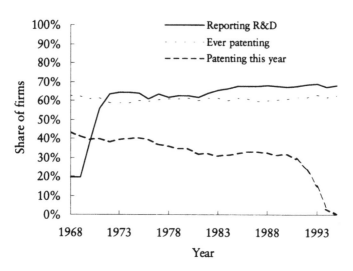

Figure 7.4 *US manufacturing – cleaned sample*

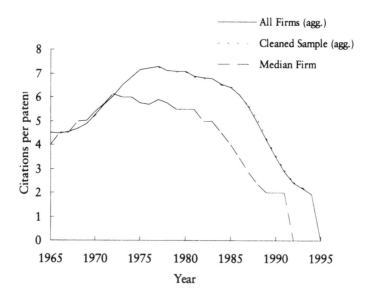

Figure 7.5 *Citations per patent by year of application*

Innovation and market value

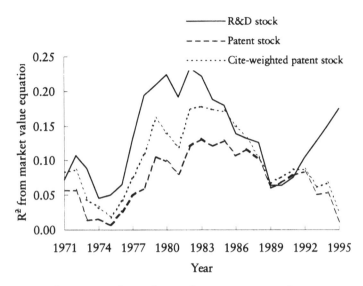

Figure 7.6 *Explaining market value with innovation stocks*

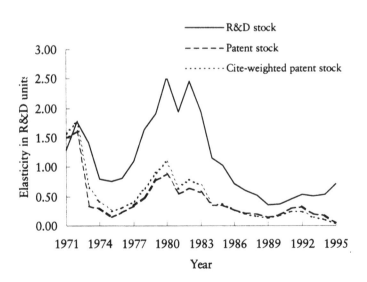

Figure 7.7 *Explaining market value with innovation stocks*

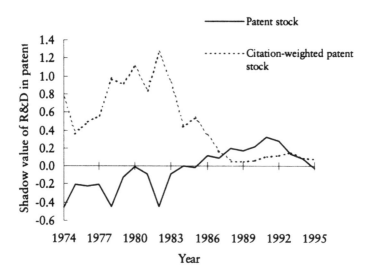

Figure 7.8 *Patent valuation coefficients in R&D units*

ning in about 1985 or earlier, because of the fact that in many cases citations are made more than ten years after the original patent is issued. These facts need to be kept in mind when looking at the estimates below.

Figure 7.6 shows the R-squares from three simple Tobin's q regressions that use different measures for the stock of innovations: the R&D stock, the stock of patents (measured at application date), and the stock of patents weighted by citations ever made. While neither patents nor citation-weighted patents do as well as R&D in explaining Tobin's q, clearly citation-weighted patents do better, especially in the earlier years where the measure is better.

Figure 7.7 shows the coefficients from this regression; in the case of patents and citation-weighted patents, the coefficients have been scaled so that they are in R&D units and can be compared with the R&D coefficient.[10] The patent measures exhibit the same decline in the 1980s as the R&D coefficients, but clearly the estimated shadow prices are much lower, which is to be expected given the way we normalised them. The main message of the figure is that the citation measures appear to work slightly better in the period where we think they are more or less complete (before 1985).

Figure 7.8 shows the coefficients that result when both patent stocks and citation-weighted patent stocks are included in the same Tobin's q equation (again measured in R&D units). The result is that the citation-weighted patent stock is clearly preferred in the pre-1985 period over the unweighted stock of patents, which has a coefficient that is zero or negative. Thus Hall, Jaffe and Trajtenberg (1998) conclude that citation-weighted measures of patents have the potential to be a more precise 'economic' measure of innovation than patents by themselves.

Conclusion

This chapter has surveyed the somewhat limited literature on the market valuation of the intangible assets associated with industrial innovation. The key findings from the recent work are the following:

1. R&D assets are valued by financial markets. A reasonable fraction of the variance in market value that remains after controlling for ordinary assets is explained by either R&D spending or the stock of R&D (with the flow coefficient averaging about 4–5 times the stock coefficient). However, there is still a fair amount of unexplained variance. This basic fact is true in the United States and the United Kingdom.
2. The R&D coefficient is not stable over time in either the United States or the United Kingdom. In the US, this coefficient reached a recent peak in the early 1980s and has declined since. This result seems to vary across industry, but industry-level findings are somewhat unstable and inconclusive.
3. Patents are informative above and beyond R&D, although the correlation is much weaker. The average R^2 for the Tobin's q–R&D relationship is approximately 0.15, while that for the patents–R&D relationship is about 0.08.
4. Citation-weighted patents are slightly more informative than patents. The average R^2 for citation-weighted patents alone is about 0.10. When both variables are included in a regression, the citation-weighted one is clearly superior.
5. However, the pattern of the patents and citation-weighted patents coefficients in the market value regression over time appears to be identical: they are measuring the same thing, but citation weights improve precision.

Many areas remain for future work in this field. In particular, the specification and estimation of the valuation equation in a setting where the assets have a very wide and skewed distribution is somewhat unsatisfactory at the moment and could be improved, both to reduce sensitivity to outliers and to increase the stability of the coefficients over time, or to model the changes in the coefficients. The patents data used here are somewhat preliminary and work can be done to improve their accuracy and coverage. The weighted and unweighted stocks of patents ought to be adjusted for changes in patenting and citation propensity over time, and the role of self-citations explored. Finally, the very interesting question of the timing of the citation effects on valuation should to be explored. But the preliminary results using this new data set show promise.

Appendix

The data used are drawn from the Compustat files and from files produced by the United States Patent Office. They include all the firms in the manufacturing sector (SIC 2000–3999) between 1976 and 1995 in a large unbalanced panel (approximately 5000 firms). The firms are all publicly traded on the New York, American, and regional stock exchanges, or traded Over-the-Counter on NASDAQ. For details on data construction, see the documentation in Hall (1990), although this data set is a new sample from a larger field than the file described there.

The chief variables from Compustat that are used are the market value of the firm at the close of the year, the book value of the physical assets, and the book value of the R&D investment. The market value is defined as the sum of the value of the common stock, the value of the preferred stock (the preferred dividends capitalised at the preferred dividend rate for medium risk companies given by Moody's), the value of the long-term debt adjusted for inflation, and the value of short-term debt net of assets. The book value is the sum of the net plant and equipment (adjusted for inflation), the inventory (adjusted for inflation), and the investments in unconsolidated subsidiaries, intangibles and others (all adjusted for inflation). Note that these intangibles are normally the goodwill and excess of market over book from acquisitions, and are not included in the R&D investment. The R&D capital stock is constructed using a declining balance formula and the past history of R&D spending with a 15 per cent depreciation rate.

The patents data have been cleaned and aggregated to the patent assignee level at the Regional Economics Institute, Case Western Reserve University. The data set matches the patent assignee names with the names of the Compustat firms and their subsidiaries in order to assign patents to each firm. This project is still underway and the current success rate of name-matching is about two-thirds, although many of the remainder firms simply do no patenting. Most of the subsidiaries are not yet included. Thus the results in this paper involving patents must be viewed as extremely preliminary, although suggestive.

Notes

1. Paper prepared for the NIESR Conference on Productivity and Competitiveness, London, 5–6 February, 1998. The new research reported in this paper is drawn from work in progress that is joint with Daehwan Kim (Harvard University), Adam Jaffe (Brandeis University and NBER), and Manual Trajtenberg (Tel Aviv University and NBER). The data construction effort was partially supported by the National Science Foundation. I am grateful to Meg Ferrando of REI, Case Western Reserve University for excellent assistance in matching the patenting data to Compustat.
2. See Fisher and McGowan (1983) for a discussion of the general problem of using accounting data for this purpose
3. Of course, it always possible to think of the additively separable form of the function as simply the first and most important terms in a more general approximation to the true function.
4. This is exactly parallel to the distinction between rate of return estimates and elasticity estimates in the productivity literature (about which many have written: see for example, Hall, 1996, for a discussion of this issue). And much the same tension exists between the two: a constant shadow value across firms is more defensible from a theoretical (market efficiency) point of view, but the constant elasticity form tends to fit the data better and be less sensitive to outliers.
5. See Hall and Kim (1997) for an exception to this rule. In that paper we use the nonlinear form of the equation explicitly.
6. Unfortunately, this quantity is difficult to work with both because it is undefined for firms that do not do R&D, and also because of its very skew distribution for those that do. Figure 7.3 presents an example that shows how much this estimate of the absolute shadow value can differ from those using the linear model when average values of the market-value knowledge–capital ratio are used to evaluate this expression.
7. The figure shows the number of firms available in each year using three different sets of criteria: the total number of firms for which we have some data, the number of firms that report R&D and have good data on capital, employment, sales and market value, and finally the number of firms that pass various data quality criteria in addition. These criteria remove firms with large jumps in their series, firms that only have tiny amounts of R&D, and a few other outliers.
8. Unlike the United States, where financial accounting standards have required the reporting of 'material' R&D expense since 1972, in the United Kingdom the requirement was imposed only in the late 1980s, so public data on R&D spend-

ing are sparse before that time. And in many continental countries, R&D reporting is not part of the standard accounting requirements even today.
9 The stock of patents is defined using a declining balance formula and a depreciation rate of 15 per cent, by analogy to the stock of R&D spending: $PS_t = 0.85 PS_{t-1} + P_t$.
10 The scaling is done in the following manner: for each year, we estimate a shadow price γ of a patent, in units of millions of dollars per patent using equation (3) with the patent stock in place of K. We convert this quantity to R&D units by dividing it by the average amount of R&D necessary to produce one patent in the sample as a whole (the ratio of R&D spending to patents for that year). We do the same computation for citations, using the average amount of R&D necessary in that year to produce one citation in the future. This is only an approximation, and a very poor one, since the meaning of citations and patents varies enormously across firms. But it does enable us to compare the estimates on the same chart.

References

Austin, D. (1993), 'An event study approach to measuring innovative output: the case of biotechnology', *American Economic Review*, 83 (May), pp. 253–8.
Ben-Zion, U. (1984), 'The R&D and investment decision and its relationship to the firm's market value: some preliminary results', in Griliches, Z. (ed.), *R&D, Patents, and Productivity*, Chicago, University of Chicago Press, pp. 299–312.
Blundell, R., Grifith, R. and van Reenen, J. (1995), 'Market share, market value, and innovation in a panel of British manufacturing firms', London, University College London Discussion Paper 95/19 (October).
Cockburn, I. and Griliches, Z. (1987), 'Industry effects and appropriability measures in the stock market's valuation of R&D and patents', *American Economic Review*, 78 (May), pp. 419–23.
Chauvin, K.W. and Hirschey, M. (1993), 'Advertising, R&D expenditures and the value of the firm', *Financial Management*, Winter, pp. 128–40.
Connolly, R.A., Hirsch, B.T. and Hirschey, M. (1986), 'Union rent seeking, intangible capital, and market value of the firm', *Review of Economics and Statistics*, 68, 4, pp. 567–77.
Connolly, R.A. and Hirschey, M. (1988), 'Market value and patents: a Bayesian approach', *Economics Letters*, 27, 1, pp. 83–7.
(1990), 'Firm size and R&D effectiveness', *Economics Letters*, 32, pp. 277–81.
Fisher, F.M. and McGowan, J.J. (1983), 'On the misuse of accounting rates of return to infer monopoly profits', *American Economic Review*, 73, 2, pp. 82–97.
Griliches, Z. (1961), 'Hedonic prices for automobiles: an econometric analysis of quality change', reprinted in Grilcihes, Z. (ed.), *Price Indexes and Quality Change: Studies in New Methods of Measurement*, Cambridge, Mass., Harvard University Press, pp. 55–87.

(1981), 'Market value, R&D, and patents', *Economics Letters*, 7, pp. 183-7.
Griliches, Z., Hall, B.H. and Pakes, A. (1991), 'R&D, patents, and market value revisited: is there a second (technological opportunity) factor?', *Economics of Innovation and New Technology*, 1, pp. 183-202.
Griliches, Z., Pakes, A. and Hall, B.H. (1987), 'The value of patents as indicators of inventive activity', in Dasgupta, P. and Stoneman, P. (eds), *Economic Policy and Technological Performance*, Cambridge, Cambridge University Press, pp. 97-124.
Hall, B.H. (1990), 'The manufacturing sector master file: 1959-1987', Cambridge, Mass., NBER Working Paper No. 3366 (May).
(1993a), 'The stock market valuation of R&D investment during the 1980s', *American Economic Review*, 83, pp. 259-64.
(1993b), 'Industrial research during the 1980s: did the rate of return fall?', *Brookings Papers on Economic Activity Micro* (2), pp. 289-344.
(1996), 'The private and social returns to research and development: what have we learned?', in Smith, B.L.R. and Barfield, C.E. (eds), *Technology, R&D and the Economy*, Washington DC, Brookings Institution and American Enterprise Institute.
Hall, B.H. and Hall, R.E. (1993), 'The value and performance of US corporations', *Brookings Papers on Economic Activity*, 2, pp. 1-50.
Hall, B.H., Jaffe, A. and Trajtenberg, M. (1998), 'The economic significance of patent citations', UC Berkeley, Nuffield College, Brandeis University, University of Tel Aviv, and NBER, work in progress.
Hall, B.H. and Kim, D. (1997), 'Valuing intangible assets: the stock market value of R&D revisited', UC Berkeley, Nuffield College, Harvard University and NBER, work in progress.
Harhoff, D., Narin, F., Scherer, F.M. and Vopel, K. (1997), 'Citation frequency and the value of patented innovation', Mannheim, ZEW Discussion Paper No. 97-27.
Hayashi, F. and Inoue, T. (1991), 'The relation between firm growth and q with multiple capital goods: theory and evidence from panel data on Japanese firms', *Econometrica*, 59, 3, pp. 731-53.
Hirschey, M. (1985), 'Market structure and market value', *Journal of Business*, 58, 1, pp. 89-98.
Hirschey, M., Richardson, V.J. and Scholz, S. (1998), 'Value relevance of nonfinancial information: the case of patent data', University of Kansas School of Business, photocopied.
Hirschey, M. and Weygandt, J.J. (1985), 'Amortization policy for advertising and research and development expenditures', *Journal of Accounting Research*, 23, pp. 326-35.
Jaffe, A. (1986), 'Technological opportunity and spillovers of R&D: evidence from firms' patents, profits, and market value', *American Economic Review*, 76, pp. 984-1001.
Lev, B.L. and Sougiannis, T. (1996), 'The capitalization, amortization, and value-relevance of R&D', *Journal of Accounting and Economics*, 21, pp. 107-38.
Mairesse, J. and Mohnen, P. (1996), 'R&D-productivity growth: what have we

learned from econometric studies?', *Proceedings of the EUNETIC Conference on Evolutionary Economics of Technological Change: Assessment of Results and New Frontiers*, Strasbourg, October 1994, pp. 817–88.

Megna, P. and Klock, M. (1993), 'The impact of intangible capital on Tobin's q in the semiconductor industry', *American Economic Review*, 83, pp. 265–9.

Shane, H. (1993), 'Patent citations as an indicator of the value of intangible assets in the semiconductor industry', The Wharton School, mimeo, November.

Stoneman, P. and Toivanen, O. (1997), 'Innovation and market value in the UK: some preliminary results', Warwick Business School, mimeo.

Trajtenberg, M. (1989), 'The welfare analysis of product innovations, with an application to computed tomography scanners', *Journal of Political Economy*, 97, 2, pp. 445–79.

(1990), 'A penny for your quotes: patent citations and the value of innovations', *Rand Journal of Economics*, 21, 1, pp. 172–87.

Waugh, F.V. (1928), 'Quality factors influencing vegetable prices, *Journal of Farm Economics*, 10, 2, pp. 185–96.

White, F.C. (1995), 'Valuation of intangible capital in agriculture', *Journal of Agricultural and Applied Economics*, 27, 2, pp. 437–45.

Wildasin, D. (1984), 'The q theory of investment with many capital goods', *American Economic Review*, 74, 1, pp. 203–10.

8 Gross investment and technological change: a diffusion based approach

PAUL STONEMAN AND MYUNG-JOONG KWON

1. Introduction

This chapter provides an overview of two other papers, Stoneman and Kwon (1998a, 1998b), which at the level of the firm and the industry respectively explore the impact of technological change (interpreted as changing technological possibilities) upon gross investment. This is a topic that has been relatively ignored in the existing investment literature, but when it has been considered is usually modelled by the addition of a technical change term into an underlying production function. In this work we take a different approach building upon the literature on the diffusion of new technologies. Although this literature (see for example Stoneman and Karshenas, 1995) is primarily concerned with the determinants of the intertemporal path of investment in a new technology, to date the investment and diffusion literatures have developed almost in isolation from each other. The methodological approach embodied in this work helps to build a bridge between the two literatures by developing a theory of gross investment upon the basis of diffusion theory and by so doing introduces technological change into the investment equation. The two approaches to investment do differ radically in their aggregative structures. The standard neoclassical investment models commonly define a capital aggregate and then analyse changes in the demand for that aggregate over time. The diffusion based approach however looks first at the demand over time for individual capital goods and then aggregates over the capital goods available. This latter aggregation enables a term in technological opportunities to enter the gross investment function.

The suggestion that technological change should impact upon gross investment has a long history; for example, Keynes (1936) in his *General Theory* discusses how changing technological opportunities will impact upon investment. Studies of the link between technological change and investment experienced a brief renaissance in the heyday of vintage growth models in the 1970s (Stoneman, 1983, provides a survey of the literature through to the early 1980s). Recently, new growth theory, as exemplified in the work of Romer (1990), has brought the topic centre stage by emphasising how investment is driven by the appearance of new technologies produced in the research sector. Lach and Schankerman (1989) have explored the link between investment and technological change using a dynamic factor demand model. Although the theoretical approach and the concept of technological opportunities used differ considerably from those used here, they find for a sample of US firms, that R&D (their measure of technological change) Granger causes investment but investment does not Granger cause R&D (in complete contrast to the early work of Schmookler, 1966, which argues for the reverse causality). Nickell and Nicolitsas (1996) have also looked at the R&D investment link using UK data. Again their approach is very different from that employed here but they find that R&D expenditure does encourage investment in most industries and that there are no positive effects in the other direction. The existing literature thus suggests that technological opportunities are a potentially important but under researched determinant of gross investment. By building upon the diffusion literature the analysis presented here provides a clearer picture of the nature of the link between technological change and investment and also utilises a much richer menu of indicators of technological change than in these earlier contributions.

The chapter is laid out as follows. In the next section we discuss some common properties of the firm level and industry level approaches. In section 3 we concentrate upon the theory and results at the level of the firm and in section 4 we concentrate upon the theory and results at the industry level. In section 5 we pull together the results and draw conclusions. In Stoneman and Kwon (1998a, 1998b) many further details of this research are documented. In this particular piece we do not 'dot all the *is* and cross all the *ts*'. To do so would take far too much space. For the full presentation the reader is referred back to the two original papers.

2 The basic approach

The two papers summarised here have a considerable amount of commonality. In each it is assumed that:
(i) Technological change encompasses new processes, products, materials and methods.
(ii) New technologies require the installation of new capital goods.
(iii) All technologies, j, are of the fixed coefficient Leontief type (with capital coefficient α_j)
(iv) The number of technologies available in time period t is $M(t)$, these technologies showing neither substitutability nor complementarity in production (technologies are additive).
(v) Gross investment is made up of two parts, adoption investment and expansion investment. Adoption investment is defined as investment undertaken to change technologies given the current level of output, expansion investment being defined as investment undertaken to expand output.
(vi) Adoption investment is modelled by using diffusion theory to generate the demand for capital good j in time t and then total adoption investment results by summing these demands over $j = 1...M(t)$.
(vii) The diffusion model used reflects (see Karshenas and Stoneman, 1993) rank, stock and epidemic effects (not order effects) with optimal adoption dates determined by the maximisation of firm present value.
(ix) The price of an individual capital good j, $P_j(t)$, is assumed to fall relative to the price of all capital goods, $P^c(t)$, the longer it has been available according to

$$P_j(t) = \phi_j P^c(t) \exp(-\Omega_j (t - T_j)) \tag{1}$$

where ϕ_j and Ω_j are (capital good specific) parameters and T_j is the date that technology j appears on the market for the first time.
(x) Expansion investment is modelled as a function of increases in output since the last peak, it being allowed that this increase may be endogenous and especially a function of technological change and/or adoption investment.
(xi) Static and dynamic versions (i.e. including a lagged dependent variable) of the models are specified and estimated.

3 The firm based model

3.1 Theory

The firm based model relies on a hazard rate formulation of an underlying diffusion model. For a given firm i (with the i subscript dropped for expositional convenience where feasible) expected adoption investment in time t at current prices, $I'(t)$, may be expressed as

$$I'(t) = \sum_j (f_j(t) P_j(t) \alpha_j q(t)) \qquad (2)$$

where $f_j(t)$ is the (unconditional) probability that the firm will install technology j in time t, $q(t)$ is the output of the firm in time t and $P_j(t)\alpha_j q(t)$ is the cost of equipping the firm in time t to produce that output.

We allow that the probability of the firm adopting technology j in time t is related to two factors. The first is an epidemic effect reflecting that the longer a technology has been on the market the more likely it is to be known to the firm and thus adopted. We measure this by $(t - T_j)$, where T_j is the date at which technology j first appeared on the market. The second factor is the net expected cost in time t of waiting until time $t + dt$ before acquiring the technology, $Y_j(t)$. Allowing that the annual gross benefit from the adoption of technology j is given by $g_j(C(t), P_k(t))$ where $C(t)$ is a vector of firm characteristics (the rank effect) and $P_k(t)$ is industry price (the stock effect) we may write that

$$Y_j(t) = r(t) P_j(t) \alpha_j q(t) - p_j(t) \alpha_j q(t) - g_j(C(t), P_k(t)) \qquad (3)$$

where $p_j(t)$ is the expected change in $P_j(t)$ in the interval $(t + dt)$ and $r(t)$ is the interest rate/discount rate.

We assume that $f_j(t)$ is related exponentially to $Y_j(t)$ and $(t - T_j)$ such that the following holds:

$$f_j(t) = \exp\{-\delta_j Y_j(t) + \phi_j(t - T_j)\} \qquad (4)$$

Through assumptions detailed in the original paper we may argue that $-Y_j(t)$ has an interior maximum whereas $(t - T_j)$ increases monotonically with time.

Assuming that $g_j(C(t), P_k(t))$ is a linear function, substituting for $P_j(t)$ from (1), defining real adoption investment as $RI'(t) = I'(t)/P^c(t)$, and summing over the technologies available in time t, $M(t)$, yields

$$\log RI'(t) = \pi_0(t) + \pi_1 \log q(t) + \pi_2 \log M(t) + \pi_3(t) P_k(t)$$
$$+ \pi_4(t)(r(t)P^c(t)q(t) - p^c(t)q(t)) + \pi_5(t)P^c(t)q(t) \quad (5)$$
$$+ C'(t)\beta_1 \delta(t)$$

where $\pi_1 = \pi_2 = 1$, and the parameters π_3 to π_5 and δ are combinations of averages across j of the basic parameters of the model and may be time varying, but π_3 is positive, π_4 and π_5 negative.

Equation (5) states that real adoption investment in firm i in time t is a function of the firm's level of output, the number of technologies available to the firm (the technological change or technological opportunities term), and a number of variables that impact upon the speed of diffusion of technology – capital prices, the expected change in capital prices (which we measure by the actual change in $P^c(t)$ between t and $t + 1$), firm characteristics and industry prices.

In order to operationalise the model we allow, because we do not have data upon the number of technologies available, that $M(t)$ can be proxied according to (6)

$$\log M(t) = \beta_4 + \beta_5 \log R(t) + \beta_6 \log \text{Pat}(t) + \beta_7 \log IRD(t) \quad (6)$$

where $R(t)$ is the firm's own R&D spend in time (t), $\text{Pat}(t)$ is the number of patents granted to the firm in time t and $IRD(t)$ is R&D in time t in the industry to which the firm belongs. In our empirical investigation we also allow for the inclusion of lagged values for these variables. We would also argue that the firm may source technology from other industries or from overseas, which sources we cannot adequately proxy. We thus also allow for firm specific effects in this relationship.

We consider three firm characteristics as relevant to the determination of the firm's return from technological acquisition, firm size (measured by employment), $EM(t)$, cash flow, $CF(t)$, and the wage rate, $w(t)$, and then specify (7). We also allow that β_{10} may be firm specific.

$$C'(t)\beta_1(t) = \beta_{10}(t) + \beta_{11}(t)EM(t) + \beta_{12}(t)CF(t) + \beta_{13}(t)w(t) \quad (7)$$

Allowing that expansion investment can be modelled by a simple additive term in $Dq_i(t)$, the increase in the firm's output since the previous peak level if positive, zero otherwise, introducing a lagged dependent variable to allow for adjustment lags, excluding the term in $P^c(t)q_i(t)$ because of multicollinearity, and defining $I_i(t)$ as the expected value of the firm's total real gross investment in time t (which we measure by actual investment plus an error term), yields (8), the equation of the dynamic model that is estimated below.

$$\log I_i(t) = b_0(t) + b_1 \log q_i(t) + b_2 \log R_i(t) + b_3 \log \text{Pat}_i(t)$$
$$+ b_4 \log IRD(t) + b_5(t) P_k(t) + b_6(t) r(t) P^c(t) q_i(t)$$
$$+ b_7(t) p^c(t) q_i(t) + b_8(t) EM_i(t) + b_9(t) CF_i(t)$$
(8)
$$+ b_{10}(t) w_i(t) + b_{11} Dq_i(t) + b_{12} \log I_i(t-1) + u_i(t)$$

where $b_0(t)$ includes the firm specific effects, $u_i(t)$ is the error term and the parameters that may be time varying are indicated as such.

3.2 Data samples and estimation methods

The sources of the data used in the econometrics below are detailed in the original paper and not repeated here. Two samples are used in the estimation:

Sample A is a panel, 1984–92, of 185 UK firms included in the 1992 DTI R&D Score-board as reporting R&D in 1992. This is a balanced panel with 1665 observations. For this sample $R_i(t)$ is measured by R&D in 1992, a non time varying covariate.

Sample B includes only those firms in sample A who in any year report R&D in that year. This is an unbalanced panel of 607 observations. For this sample $R_i(t)$ is measured by actual R&D in time t. This sample, although smaller than sample A, has the advantage of providing a better measure of $R_i(t)$.

We tested for non-stationarity in the data using a unit root test (see Im, Pesaran and Shin, 1995) and found that there is stationarity (we are grateful to Dr Shin for providing the unit root test program).

We have estimated both static and dynamic versions of the model, the static version being as (8) but with the term in $\log I_i(t-1)$ omitted. Given the inclusion of firm specific effects we estimate the dynamic model on sample A using GMM with one period lagged variables as instruments. We estimate the static model on sample A using GLS. The nature of the sample precludes us from using GMM to estimate the dynamic version of the model using sample B, we thus estimate only the static model on this sample. After a Hausman test, for sample B we have chosen to present the results using a Within Group Estimator.

In these estimates we have taken account of the fact that Dq may be a function of past investment and other variables already incorporated in the model (i.e. may itself be a function of technological change) and have thus replaced Dq by the predictions of a regression of Dq on current and lagged values of the other independent variables in the model. One should also note that potentially, given the theory, a number of the parameters of the model may be time varying. We have tested for this but find no evidence for such time variation.

3.3 Empirical results

In table 8.1 below we present our estimates with heteroscedastic consistent t statistics in parentheses and significance at 5 per cent indicated by **, at 10 per cent by *. The diagnostic statistics for the estimates tabulated are good. A Wald test rejects at the 1 per cent level the hypothesis that all coefficients are zero for each estimate. LM tests show that there is no second order serial correlation in the GMM estimates, thus confirming the validity of the instruments used.

Introducing lagged values for log R_i, log IRD and log Pat_i into the specification made little difference to the estimates and the lagged values for these variables were not significant. The hypothesis of no lagged effects cannot be rejected. Although *a priori* one might expect that past values of these variables would be included in any proxy for $M(t)$, it is often found in data of this kind that current levels of R&D and patenting are good proxies for past R&D and patenting and this may well rationalise the result. Experiments were undertaken with the inclusion of time dummies but these were highly correlated with the IRD variable and were thus not pursued.

Consider the preferred GMM estimates first. These estimates use the balanced panel, include both firm specific effects and a lagged dependent variable and allow for the potential endogeneity of the right hand side variables. Our tests reject the hypothesis that there are no firm specific effects. The lagged investment variable is highly significant, implying that there are dynamic effects in the determination of gross investment.

Log $q_i(t)$ is significant and a long run coefficient estimate of 0.60 is indicated. This is less than the value of unity that the theory suggests but, of necessity, $q_i(t)$ has been measured as real sales whereas the theoretical construct is the number of units of output produced. We thus do not consider the divergence from unity to be a major problem.

Given that in the balanced sample $R_i(t)$ is a non time varying covariate the differencing involved in GMM estimation removes the variable. Technological opportunity is thus represented by the Patenting and Industry R&D variables. The latter is significant with a coefficient of the expected sign, the former carries a negative coefficient but is not

Table 8.1 *Regression estimates (firm level)*

Variable	Balanced panel GLS	Balanced panel GMM	Unbalanced panel WG
Constant	-5.99**	–	–
	(-11.33)		
$\log I_i(t-1)$	–	0.29**	–
		(6.58)	
$\log q_i(t)$	0.89**	0.43**	0.10**
	(38.01)	(5.51)	(2.84)
$\log R_i(t)$	0.088**	–	0.13**
	(4.07)		(4.82)
$\log IRD(t)$	0.19**	0.26**	0.24*
	(5.42)	(3.20)	(1.90)
$\log Pat_i(t)$	0.0001	-0.01	0.00006
	(0.02)	(-1.15)	(0.01)
$r(t)P^c(t)q_i(t)$	-0.22**	-2.99*	0.06
	(-3.10)	(-1.65)	(1.22)
$p^c(t)q_i(t)$	0.0003	0.004	-0.0001
	(1.50)	(1.06)	(-0.38)
$P_k(t)$	23.60**	23.35**	3.93
	(5.85)	(3.51)	(1.10)
$EM_i(t)$	0.004*	0.06	0.011**
	(1.85)	(0.05)	(3.05)
$CF_i(t)$	0.11**	-0.77	-0.03
	(3.22)	(-0.47)	(-1.18)
$w_i(t)$	0.49	-2.84	-3.84
	(0.63)	(-1.02)	(-1.08)
$Dq_i(t)$	0.02**	0.04**	1.57
	(2.91)	(3.37)	(1.44)
Sargan Test for instrument validity	–	175.7 (df = 95)	–
LM test for first order serial correlation	0.11	-14.13	1.46
LM test for second order serial correlation	–	0.19	–
Wald test for all coefficients = 0	666 (df=12)	250.78 (df=11)	783 (df=187)
Wald test for technology opportunities coefficients = 0 ($b_2 = b_3 = b_4 = 0$)	36.1 (df = 3)	11.33 (df = 2)	52.1 (df = 3)
Wald test for diffusion coefficients = 0 ($b_5 = b_6 = b_7 = b_8 = b_9 = b_{10} = 0$)	66.4 (df = 6)	18.58 (df = 6)	14.2 (df = 6)
Number of observations	1665	1395	697

significant. A Wald test rejects the hypothesis that the coefficients on all technological opportunities variables are zero ($b_3 = b_4 = 0$).

The cost of acquiring new technology $r(t)\ P^c(t)\ q_i(t)$ is significant with a coefficient of the expected sign; however the expected change in the cost of acquisition, $p^c(t)\ q_i(t)$, is not significant although of the correct sign (this is consistent with other empirical work on the role of price expectations in the diffusion process, see Stoneman and Kwon, 1994, 1996).

The industry price variable indicating the impact of demand changes and possibly previous adoption is also significant, carrying a coefficient of the expected positive sign. Similarly the expansion investment term is also significant with the expected positive sign. However none of the firm characteristics variables is individually significant. These are the variables that determine the differences between firms in the return to adoption of new technologies and thus the different dates of adoption. However, (i) we cannot reject the hypothesis of firm specific effects which may be picking up some of the impact of these characteristics variables and (ii) the Wald test rejects the hypothesis that all the diffusion related coefficients are zero ($b_5 = b_6 = b_7 = b_8 = b_9 = b_{10} = 0$).

As stated above, the first differencing involved in the GMM estimation precludes the estimation of a coefficient for $R_j(t)$. We have thus also estimated a static model for the balanced panel. These GLS estimates indicate that firm R&D has a significant and positive impact on real gross investment. Compared to the GMM results the impact of industry R&D is reduced (as one would expect); however, the coefficient on patenting is still insignificant although now positive (as expected *a priori*). The balanced panel data thus indicate that firm and industry level R&D as indicators of technological opportunity have positive and significant impacts on firm level gross investment.

The balanced panel GLS results also confirm the GMM results relating to the signs and significance of the coefficient on firm output (and in fact the coefficient estimate is closer to the theoretically pure prediction of unity), the costs of adoption (where with GLS the coefficient is smaller but more significant), industry price and the expansion investment term. In the GLS estimates we also see that certain firm characteristics variables (employment and cash flow) are individually significant with coefficients of the expected sign. We cannot reject the hypothesis of firm specific effects. Wald tests also indicate that we can reject the joint hypothesis that all technological opportunity and all diffusion related variables have zero coefficients.

The Within Group estimators of the static model using the unbalanced panel data set largely confirm the results from the balanced panel. Firm

Table 8.2 *Firm level elasticity estimates*

Variable	Elasticity
$q_i(t)$	0.60
$IRD(t)$	0.37
$r(t)\,P^c(t)\,q_i(t)$	−0.78
$P_k(t)$	0.14
$Dq_i(t)$	0.04

output has a significant and positive effect on gross investment (although the coefficient is small). Industry and firm level R&D also have positive and significant effects, as in the GLS estimates on the balanced panel. Patents are not significant. However, in these estimates neither of the price variables is individually significant nor is the output price variable. Of the firm characteristics only firm size (with a positive coefficient) is significant. The hypothesis of firm specific effects again cannot be rejected. The expansion investment term is not significant. However Wald tests again enable us to reject the hypotheses that all diffusion related coefficients are zero.

One may conclude from these various estimates that, at the firm level at least, there is considerable support for this approach to the modelling of gross investment. In particular the estimates indicate that both technological opportunities variables and diffusion related variables have a significant effect on gross investment. In table 8.2 we present long run elasticities (calculated at the sample means) of $I_i(t)$ with respect to each of the significant independent variables in the GMM estimates of the balanced sample.

These elasticity estimates indicate that the variables reflecting technological opportunity and the costs and benefits of technology adoption are quantitatively significant.

4 The industry model

The industry based model differs from the firm based model in several small but quite important dimensions. First, the formulation is no longer probabilistic; second, we do not include an epidemic effect which in any case would be incorporated in to the constant term in the estimating equation as in the firm based model; third, we take the opportunity of further work to refine the modelling of adoption investment; and finally, we also have taken the opportunity to consider and deal more adequately with issues relating to the endogeneity of regressors and multicollinearity between the technological change proxies and industry output.

4.1 Theory

The basis of the industry model is to first identify, for each industry k, a critical or threshold level of output in time t, $q_{jk}{}^*(t)$, above which a firm in that industry will be a user of technology j and below which it will wait before acquisition of this technology. Given the firm size distribution in the industry, reductions in $q_{jk}{}^*(t)$ over time will trace out the path of the proportion of industry output produced on technology j, the diffusion path of technology j, from which one may derive gross investment in technology j at each time t. Summing over the technologies available, $j = 1...M(t)$, one may then derive total gross investment in new technologies.

Defining $F(q_{jk}{}^*(t))$ as the proportion of output in industry k produced by firms of a size greater than $q_{jk}{}^*(t)$, $Q_k(t)$ as the output of industry k in time t, D as the differential operator and α_j and $P_j(t)$ as in the firm level model, adoption investment in technology j in industry k in time t at current prices, $I'_{jk}(t)$ is given by

$$I'_{jk}(t) = \alpha_j P_j(t) Q_k(t) DF(q_{jk}{}^*(t)) \tag{9}$$

Using a similar theoretical procedure as in the firm level model enables us to derive that threshold level of output for technology j is given by (10)

$$q_{jk}{}^*(t) = G(r(t)P^c(t), p^c(t), P_k(t), C_k(t), t - T_j) \tag{10}$$

where $P^c(t)$ is the price of capital goods, $r(t)$ the discount rate/interest rate, $p^c(t)$ the change in the price of capital goods, $P_k(t)$ industry price, $C_k(t)$ a vector of industry characteristics, and $(t - T_j)$ the time for which technology j has been on the market (all at time t). After similar aggregations to those in the firm level model above we generate that real adoption investment is given by

$$\log RI_k(t) = \pi_{0k}(t) + \log Q_k(t) + \log M_k(t) + \pi_1(t) D(r(t) P^c(t))$$
$$+ \pi_2(t) Dp^c(t) + \pi_3(t) DP_k(t) + \pi_4(t) DC_k(t) \tag{11}$$

where $\log Q_k(t)$ and $\log M_k(t)$ carry unit parameters and other parameters, which are essentially averages across j of the basic parameters of the model, are potentially time varying. Equation (11) states that real adoption investment in an industry depends upon industry output, the number of technologies available for adoption (the technological change or technological opportunities variable) and a number of variables that reflect changes in the average cost and benefits of adoption (changes in the prices of capital goods, expected changes in the price of capital goods, industry characteristics and industry price). It is worth noting however that because of the way this version of the model is constructed, it is changes in the benefits of adoption that matter rather than the level of the benefits of adoption (as found in the firm level model).

Expansion investment is modelled as follows. We argue that real expansion investment, $E(t)$ will be linearly related to the increase in industry output since its previous peak. Defining $DQ_k(t)$ as this increase if positive or otherwise zero, we write that

$$E_k(t) = \tau DQ_k(t) \tag{12}$$

where τ is a parameter, $\tau > 0$. Defining total real gross investment, $I_k(t)$, as real adoption investment plus real expansion investment, defining $RI_k(t) = \tau' Q_k(t)$, and using the approximation that $\log(1 + x) = x$, we may write that

$$\log I_k(t) = \log RI_k(t) + (\tau / \tau')(DQ_k(t) / Q_k(t)) \tag{13}$$

where we interpret τ / τ' as a parameter (which may be time varying) and $DQ_k(t)/Q_k(t) = G_k(t)$ as the rate of growth of industry output when positive and otherwise zero.

In order to operationalise the model we allow that the number of technologies available can be proxied as in (14)

$$\log M_k(t) = \mu_0 + \mu_1 \log IRD_k(t) + \mu_2 \log IPT_k(t) + \mu_3 \log TR_k(t) \\ + \mu_4 \log IP_k(t) + \mu_5 \log IU_k(t) \tag{14}$$

where, taking advantage of the wider choice of technology indicators available at the industry level, $IRD_k(t)$ is R&D in industry k in time t, $IPT_k(t)$ is the number of patents granted to firms in industry k in time t, $TR_k(t)$ is the number of trademarks newly registered to firms in industry k in time t and $IP_k(t)$ and $IU_k(t)$ are the number of innovations respectively produced and used by firms in industry k in time t.

As industry characteristics we include the industry wage, $W_k(t)$, the industry concentration ratio $CR_k(t)$ and the number of firms in the industry $NF_k(t)$ as relevant as per (15).

$$DC_k(t) = \beta_0 + \beta_1 DW_k(t) + \beta_2 DCR_k(t) + \beta_3 DNF_k(t) \qquad (15)$$

Finally, because of (i) potential endogeneity and (ii) a high degree of multicollinearity in the data set between $Q_k(t)$ and other regressors (especially the technology indicators) we also substitute for log $Q_k(t)$, allowing that industry output is a function of industry price and the technological state of the industry summarised by $M_k(t)$ as in (16).

$$\log Q_k(t) = \beta_4 + \beta_5 \log P_k(t) + \beta_6 \log M_k(t) \qquad (16)$$

where we expect $\beta_5 < 0$ and $\beta_6 > 0$.

After substitution for $Q_k(t)$, $M_k(t)$ and $C_k(t)$ and the addition of a lagged dependent variable we generate the estimating equation of the industry model (17).

$$\begin{aligned}\log I_k(t) = & b_0(t) + b_1 \log P_k(t) + b_2 \log RD_k(t) \\ & + b_3 \log PT_k(t) + b_4 \log TR_k(t) + b_5 \log IP_k(t) \\ & + b_6 \log IU_k(t) + b_7(t) D(r(t) P^c(t)) + b_8(t) Dp^c(t) \\ & + b_9(t) DP_k(t) + b_{10}(t) DW_k(t) + b_{11}(t) DCR_k(t) \\ & + b_{12}(t) DNF_k(t) + b_{13} G_k(t) + b_{14} \log I_k(t-1).\end{aligned} \qquad (17)$$

where indicated parameters may again be time varying.

4.2 Data and estimation methods

The data for the estimation of the industry model (equation (17)) are a panel, 1968–90, of 23 three digit UK industries. To save space we again omit details re sources and sample characteristics which are detailed in the original paper. The data on innovations used and produced are only available for the 1968–93 period. We thus define a sub-sample covering just this period (with 322 observations) labelled sample A, labelling the whole sample (with 483 observations) as sample B. We tested for non-stationarity in the data and found that there is stationarity.

We have estimated static and dynamic versions of the model but here report only the estimates of the dynamic model which anyway nests the static model. The dynamic model, with industry specific effects, is estimated using a GMM estimator with three period lagged values used as instruments. A Sargan test accepts the validity of these instruments at the 6 per cent significance level. $G_k(t)$ has again been replaced with the predictions from a regression of $G_k(t)$ on the other regressors. In addition, as it did not perform at all well, the patents variable has been dropped (it was never significant but its inclusion slightly reduces the significance of the R&D variable). We also entered $RD_k(t)$, $TR_k(t)$, $IP_k(t)$ and $IU_k(t)$ with one period lags to help overcome any remaining endogeneity problems. Time dummies have also been introduced and these are found to be jointly significant. The GMM estimates are presented with a constant included but the constant is never significant. As in the firm level model we tested whether there was any evidence of changes in the parameters over time but no such evidence was found.

4.3 Empirical results

In table 8.3 we present the GMM estimates of the dynamic model for the two samples. Significance at the 5 per cent level is indicated by ** and at the 10 per cent level by * with heteroscedastic consistent t statistics presented in parentheses. The diagnostic indicators are good in that using Wald tests we can reject the hypotheses that (i) jointly all coefficients are zero (ii) that the time dummies are not significant and (iii) that the instruments are invalid.

For both samples the lagged dependent variable is highly significant and thus the dynamic model is to be preferred to the static model. Log $P_k(t)$ is also significant at the 10 per cent level for both samples and carries the expected negative coefficient. It should be recalled that log $P_k(t)$ was entered to represent log $Q_k(t)$ and thus picks up the effect of industry output on gross investment.

The variables representing technological opportunity are log $RD_k(t)$, log $TR_k(t)$, log $IP_k(t)$ and log $IU_k(t)$. In sample A both industry R&D and the innovations produced variables are significant at the 5 per cent level. For sample B where the innovation variables are excluded both industry R&D (at 5 per cent) and trademarks registered (at 10 per cent) are significant. In each case the variables carry the expected positive coefficients. Although the innovations variables through (16) are partly acting as determinants of $Q_k(t)$, it appears reasonable to suggest that these results confirm that technological opportunity impacts positively on industry gross investment. A Wald test rejects the hypothesis that the technological opportunities variables jointly have no impact on gross investment.

Of those variables that we have incorporated as determinants of the speed of diffusion, the change in the rental cost of capital $D(r(t).P^c(t))$ is significant at the 10 per cent level in sample A but is not significant for sample B. It also carries the wrong sign in both samples. However $Dp^c(t)$ (the change in the expected change in the price of capital) is significant at the 10 per cent level in both samples and carries the expected sign. The

Table 8.3 Regression estimates (industry level)

Period	1968–83		1968–90	
Constant	−0.03	(−0.61)	−0.0047	(−0.08)
$\log I_k(t-1)$	0.71	(15.07)**	0.8	(26.6)**
$\log P_k(t)$	−0.75	(−2.01)**	−0.69	(−2.04)**
$\log RD_k(t)$	0.013	(1.92)**	0.029	(2.78)**
$\log TR_k(t)$	0.04	(0.39)	0.136	(1.66)*
$\log IP_k(t)$	0.013	(3.52)**	–	
$\log IU_k(t)$	0.001	(0.44)	–	
$D(r(t).P^c(t))$	0.073	(1.76)*	0.059	(1.06)
$Dp^c(t)$	0.23	(1.64)*	0.3	(1.88)*
$DP_k(t)$	0.002	(2.4)**	0.003	(3.61)**
$D\bar{W}_k(t)$	0.0003	(0.26)	−0.001	(−1.42)
$DCR_k(t)$	0.003	(0.89)	0.0003	(0.1)
$DNF_k(t)$	0.000013	(0.47)	0.000028	(1.28)
$G_k(t)$	0.73	(4.27)**	1.14	(6.24)**
Wald test (χ^2) for significance of time dummies	36.96		88.1	
Sargan test for instrument validity	221.32		247.2	
Wald test for $b_2=b_3=b_4=b_5=b_6=0$	16.5		44.3	
Wald test for $b_7=b_8=b_9=b_{10}=b_{11}=b_{12}=0$	12.06		20.23	
Wald test for all all coefficients = 0	335.48		985.9	
No of observations	322		483	

change in industry price, $DP_k(t)$, carries the expected positive coefficient and is also significant at the 5 per cent level for both samples. However, the impacts of the industry characteristics variables are very imprecisely estimated, none of the three being significant, although a Wald test enables us to reject the hypothesis that jointly the diffusion related variables do not affect gross investment.

Finally, as stated above, expansion investment is included in this model through the term in $G_k(t)$, measuring the growth of industry output. This is seen as an improvement to the firm level model. $G_k(t)$ carries the expected positive coefficient which is also significant at the 5 per cent level.

These results thus indicate that again this is a valid approach to modelling gross investment. The particular estimates confirm that technological opportunity affects gross investment and that those variables that will affect the diffusion speed (largely) behave as expected. The improved modelling of expansion investment is also successful. Again the dynamic version of the model with its assumed adjustment lags is to be preferred to a static formulation.

5 Discussion and conclusions

Existing literature, especially the empirical literature, has largely ignored the role that technological change might play in investment determination, although it does carry an undercurrent implying that it is reasonable to argue that technological change does have a role. Here, in contrast, we place technological change at the heart of the modelling of gross investment.

Two pieces of research, one at the firm level and one at the industry level, are discussed. These two pieces have common antecedents and approaches. In both, two types of gross investment are defined, adoption investment and expansion investment, it being argued that these are in turn functions of technological change, the former directly the latter indirectly. The latter is modelled in a standard way as a function of output change or growth. The former is modelled utilising the theory of technological diffusion, providing a unique combination of two related fields of study that have previously developed independently.

Adoption investment is modelled by allowing that firms are being offered over time a changing number of technologies in which they may invest. These technologies are assumed additive. Theoretically an optimal adoption date for each technology by each firm is derived, aggregation across technologies yields an expression for total adoption investment by the firm and a further aggregation across firms yields total adoption investment by the industry.

It is predicted that gross adoption investment will increase with (i) the number of technologies available (ii) the diffusion speed and (iii) the amount of output to be equipped with new technology.

The models are estimated on panel data sets covering UK manufacturing industries. The empirical results confirm that: the proxies used to measure technological opportunities (R&D, patents issued, trademarks registered and innovation counts) have a positive and significant effect on gross investment; those factors that determine the speed of diffusion largely impact as predicted; both adoption and expansion investment can be identified as having separate effects on gross investment; there are adjustment lags in the process.

Our estimates are sufficiently well defined to enable us to argue that technology effects are quantitatively important and thus that models of investment that exclude technological opportunities may well suffer from mis-specification. Of more importance however, and perhaps this is our major finding, the empirical estimates are consistent with the view that gross investment may be successfully modelled via an approach built upon the foundations of the analysis of technological diffusion. This approach does much more than just add a technological change or opportunities term into a standard investment function. Technological change is placed at the heart of the model of investment and as such the role of technological change is endemic to the model rather than additional.

Viewing gross investment as (at least in part) reflecting the process of new technology diffusion provides a route by which one may also approach the (micro) economics of investment policy. Although to date the theory and empirics of diffusion policy have not merited extensive discussion (see Stoneman and Diederen, 1994, for an accessible survey), the extant literature does provide a number of insights into why a free market economy may underinvest in the diffusion of new technology and thus also provides a market failure rationale for government policies to stimulate gross investment. (Of course the rationale is not itself a sufficient reason for intervention; it must also be shown that intervention will be effective.) The policy literature extends beyond the diffusion models used here, and *inter alia* suggests the following factors that may lead to suboptimal diffusion speeds: information deficiencies; information and other non-appropriable externalities; market power effects in capital good supply; incomplete insurance markets; and inefficient capital markets. Such factors provide a justification of, for example, tax incentives to investment, risk sharing policies or information provision policies. The realisation that technological change plays a major role in the gross investment process thus provides a rationale for policy intervention that is much more soundly based than the view that commonly underlies arguments in support of stimulating gross investment, i.e., that more would be better.

References

Im, K.S., Pesaran, M. and Shin, Y. (1995), 'Testing for unit roots in heterogeneous panels', DAE Working Paper series No. 9526, Cambridge University.

Karshenas, M. and Stoneman, P. (1993), 'Rank, stock, order and epidemic effects in the diffusion of new process technology', *Rand Journal of Economics*, 24, 4, pp. 503–28.

Keynes, J.M. (1936), *The General Theory of Employment Interest and Money*,

London, Macmillan.

Lach, S. and Schankerman, M. (1989), 'Dynamics of R&D and investment in the scientific sector', *Journal of Political Economy*, 97, 4, pp. 880–904.

Nickell, S. and Nicolitsas, D. (1996), 'Does innovation encourage investment in fixed capital?', Discussion Paper No. 309, Centre for Economic Performance, LSE, October.

Romer, P. (1990), 'Endogenous technological change', *Journal of Political Economy*, 98, 5, Part 2, pp. 571–602

Schmookler, J. (1966), *Invention and Economic Growth*, Cambridge, Mass., Harvard University Press.

Stoneman, P. (1983), *The Economic Analysis of Technological Change*, Oxford, Oxford University Press

Stoneman, P. and Diederen, P. (1994), 'Technology diffusion and public policy', *Economic Journal*, 104, pp. 918–30.

Stoneman, P.and Karshenas, M. (1995), 'Technological diffusion', in Stoneman, P. (ed.), *Handbook of the Economics of Innovation and Technological Change*, Oxford, Blackwell.

Stoneman, P.and Kwon, M. J. (1994), 'The diffusion of multiple technologies', *Economic Journal*, 164, pp. 420–31.

(1996), 'The impact of technology adoption on firm profitability', *Economic Journal*, 107, pp. 952–62.

(1998a), 'Gross investment and technological change', *Economics of Innovation and New Technology*, 7, 3, pp. 221–43.

(1998b), 'Technological opportunity, technological diffusion and gross investment: an inter industry approach', mimeo, University of Warwick, February.

9 National systems of innovation under strain: the internationalisation of corporate R&D

PARI PATEL AND KEITH PAVITT[1]

1 National systems of innovation

We shall explore below the growing strain between national systems of innovation, on the one hand, and the internationalisation of corporate innovative activities, on the other. What national systems of innovation encompass and what they actually do is still subject to varying interpretations and debate (Lundvall, 1992, Nelson, 1993, Patel and Pavitt, 1994, Freeman, 1995, Edquist, 1997). In broad terms, they can be defined in terms of the institutions involved in the generation, commercialisation and diffusion of new and better products, processes and services (i.e. technical change), and of the incentive structures and competencies in these institutions that influence the rate and direction of such change. In the context of this volume's primary concern with productivity, competitiveness and technical change, we use a more restricted definition, namely, the national investments in the knowledge-generating activities that are a necessary complement to investments in equipment, in increasing efficiency and in maintaining or increasing competitiveness. Recent research on why national growth rates differ identifies two essential components of such knowledge-generating activities: education and training in all countries, and R&D in the industrially advanced countries (Fagerberg, 1994). Here, we shall concentrate on the latter.[2]

One essential element of contemporary knowledge generation[3] is its specialised nature.

- Specialisation by discipline within science and technology.
- Specialisation by corporate function inside the business firm, with the establishment of R&D laboratories; and – within the corporate R&D

function – specialisation between the development function concerned with product and process development, and the research function exploring options for future product development.
- Specialisation by institution within countries, with R&D laboratories funded by companies, and by governments – either directly or through universities and similar organisations.

Specialisation implies co-ordination – and even integration – so that an essential feature of all effective systems of innovation are the linkages (networks) between their component parts: between disciplines, between corporate functions and between institutions. Our concern here is institutional linkages – between corporate R&D, on the one hand, and the publicly funded science base in universities and similar institutions, on the other. In this context, recent empirical research confirms the existence of *national* systems of innovation. There is a strong national bias in linkages between business practitioners and academic research (Hicks, Izard and Martin, 1996, Narin, Hamilton and Olivastro, 1997). This reflects the linguistic and geographic constraints imposed by person-embodied exchanges and transfers of tacit knowledge.[4] As a consequence, high quality basic research in an OECD country (measured as citations per published paper) is correlated with high level technology (measured as business-funded R&D as a share of GDP) (Patel and Pavitt, 1994, Kealey, 1996). In addition, Arundel, van de Paal and Soete (1995) have shown that non-market corporate linkages to 'public knowledge' are in fact much more difficult to forge across national boundaries than within them.

It is in this context that we shall now examine the implications of the globalisation of corporate activities for the so far privileged links between the national science base and national innovative activities. Scholarly opinion on what is in fact happening is divided.[5] We shall first present the facts of the case, before identify underlying causal factors and implications for national and corporate policies.

2 The internationalisation of corporate R&D

Most studies concerned with analysing the internationalisation of corporate R&D are based on one of two sets of measures:

- *R&D expenditures and employees*, where the OECD (1997) has recently brought together evidence from national surveys on the shares of domestic business funded R&D performed by foreign firms, and

of R&D funded by domestically owned firms that is performed outside their home country.
- Patent statistics (Etemad and Seguin-Dulude, 1987, Cantwell, 1992, Patel and Pavitt, 1991, Patel, 1995, 1996, Patel and Vega, 1998, 1999), where the inventor's address given in each published patent, is used as a proxy measure for the geographical location of R&D activities.

This section is based mainly on the results of our own research based on firm-level patent data. The validity of this measure has been extensively discussed elsewhere.[6] Suffice to say that patenting-based data can be analysed in much greater detail and with much greater consistency than the available data on R&D activities. Moreover, as we shall now see, the patterns revealed by these patenting statistics are consistent with those revealed by the R&D statistics that are available.[7]

2.1 Continuing reliance on the home country

Our most recent research (Patel and Vega, 1999) is based on a systematic analysis of 359 of the world's largest companies[8] (from the Fortune 500 list) that were technologically active in the 1990s. Table 9.1 shows that the patenting and the R&D data tell the same story.[9] Firms continue to perform a high proportion of their innovative activities in their home countries.[10] Japanese firms have the least globalised structure of innovative activities, and European firms the most. Within Europe, the share of corporate technological activities performed outside the home country is higher for firms from small countries (more than 50 per cent in those from Belgium, Netherlands and Switzerland) than in those from large countries (a third or less in firms from France, Germany and Italy). But there are exceptions, with large firms from the UK performing more than half outside the UK and Finnish firms performing only about a quarter outside Finland.

Between the early 1980s and the mid-1990s, large firms increased the proportion of their innovative activities performed outside their home country by a modest 2.4 percentage points. The increases for European firms were larger, but the Japanese firms show a decrease. Within Europe, increases were highest for French firms, who expanded their activities more in other European countries than the US (Patel and Vega, 1999). Our earlier analysis showed that most of the increases in foreign shares have been consequences of foreign acquisitions, rather than of an international redeployment of R&D activities (Patel, 1995).

Table 9.1 *Internationalisation of corporate technology*

Nationality	% share of US patents in 1992–96		% share of R&D exp. abroad	Change in % abroad (US patents) since 1980–84
	home	abroad		
Japan	97.4	2.6	2.1 (1993)	−0.7
US	92.0	8.0	11.9 (1994)	2.2
Europe	77.3	22.7*		3.3
Belgium	33.2	66.8		4.9
Finland	71.2	28.8	24.0 (1992)	6.0
France	65.4	34.6		12.9
Germany	78.2	21.8	18.0 (1995)	6.4
Italy	77.9	22.1		7.4
Netherlands	40.1	59.9		6.6
Sweden	64.0	36.0	21.8 (1995)	−5.7
Switzerland	42.0	58.0		8.2
UK	47.6	52.4		7.6
All firms	87.4	12.6	11.0 (1997)	2.4

Source for R&D data: OECD (1997), EC (1997).
Note: *The proportion of total activities for all the European countries listed in this table located outside Europe.

Contrary to a widely held view, the degree of internationalisation of R&D is not positively associated with high technology. Table 9.2 shows that, with the notable exception of pharmaceuticals (and to a lesser extent chemicals), the proportion of firms' innovative activities performed domestically *increases* with the technology intensity of the industry. Our previous analysis (Patel, 1996) showed that this relationship also holds at the level of the firm. This reflects the influence of the following factors:

- at the industry level, the need to adapt 'traditional' products to local tastes (e.g. food and drink, building materials), and to locate technological activities close to available raw materials (e.g. petroleum, food and drink, building materials);
- at the industry level, the smaller need to adapt high-technology products (e.g. civil aircraft, automobiles) to local requirements;
- at the industry and the firm level, (a) the positive external economies of links with the local science base and supply of skills, sources of finance, and local suppliers and customers (Vernon, 1960, 1966);

Table 9.2 *Comparison of large firms' patenting at home, and their R&D intensity, by product group*

Principal product group	% of patenting in home country (1992–6)	R&D intensity* (1992)
Food, drink and tobacco	56.3	1.0
Rubber & plastics	71.4	2.5
Pharmaceuticals	78.3	10.4
Chemicals	78.5	4.5
Mining & petroleum	80.4	0.7
Building materials	83.2	1.8
Metals	87.7	1.3
Electrical/electronics	87.7	6.1
Machinery	88.5	2.3
Computers	92.5	7.0
Motor vehicles and parts	93.6	4.0
Paper	93.7	1.0
Aerospace	94.1	6.9
Photography and photocopy	95.0	5.9
All firms	12.6	4.0

Note: *R&D as a proportion of sales.

(b) the efficiency gains within firms from the close co-ordination of functional activities, and the integration of tacit knowledge, necessary for the launching of major innovations (Rothwell, 1977, Patel and Pavitt, 1991).

2.2 Location of foreign corporate R&D: 'triadisation' not 'globalisation'

Less than 1 per cent of these firms' foreign innovative activities are located outside the 'triad' countries, showing that the process of internationalisation of technological activities can at best be described as 'triadisation' rather than 'globalisation'. Within the triad countries, Japan is the least favoured foreign location, and the US, Germany and the UK the most favoured, together accounting for around 70 per cent of all foreign technological activities. In our previous research (Patel 1995), based on a larger population of firms (around 600), we showed that about 60 per cent of the firms had no foreign technological activity at the end of the 1980s, about a quarter were active in one or two foreign countries, and only about 15 per cent in more than two.

Table 9.3 *Importance of foreign large firms in national technology, 1992–6*

% of national totals	Foreign controlled US patenting (1992–6)	Foreign controlled R&D	Foreign controlled production (1994)
Japan	1.1	1.3 (1991)	2.8
US	4.0	11.3 (1994)	15.5
Europe	12.4*		
Austria	12.5		25.7**
Belgium	53.6		
Finland	3.7	7.3 (1995)	7.6
France	11.3	13.0 (1992)	21.0
Germany	9.6	14.1 (1993)	28.1
Italy	10.0		
Netherlands	13.2	14.8 (1995)	42.4
Sweden	13.6	11.2 (1994)	18.7
Switzerland	5.8		
UK	20.3	15.0 (1994)	22.3

Source: OECD (1997) and Patel and Vega (1999).
Notes: *Foreign here refers to the share of all non-European firms. **1991.

2.3 Large foreign firms in host countries' innovative activities.

Table 9.3 shows that foreign large firms are relatively much more important sources of innovative activities in Europe than in either Japan or the US.[11] The share of foreign large firms in national innovative activities in most European countries lies between 10 and 20 per cent, with Finland and Switzerland as outliers with substantially less than 10 per cent.

Table 9.3 also shows that in general foreign firms account for a larger proportion of industrial production than of technology (measured in terms of both patenting and R&D), thereby confirming that on average foreign firms have a lower technology intensity than domestic firms.[12]

2.4 Domestic and foreign firms in national systems of innovation

Table 9.3 showed that, amongst the larger R&D spending countries, the UK has a higher proportion of national technological activities performed by foreign firms. Table 9.4 compares the UK with other major countries in more detail in seven broad technological fields covering all technologi-

Table 9.4 *Relative importance of national and foreign large firms,[a] by broad technical field, 1992–6*

% of national totals	Japan Nat.	Japan For.	USA Nat.	USA For.	France Nat.	France For.	Germany Nat.	Germany For.	UK Nat.	UK For.
Electrical & electronics	73.4	0.9	39.3	3.9	40.4	13.1	47.8	17.2	14.4	33.0
Chemicals & pharma.	45.6	1.9	39.1	9.1	42.6	15.2	71.8	7.2	36.8	24.4
Process technology	53.8	1.8	32.6	6.3	51.6	7.9	57.2	8.8	33.6	17.0
Automobiles	81.1	0.4	36.0	3.5	22.0	15.0	63.4	3.0	30.7	22.1
Machinery & aircraft	40.4	0.8	14.5	1.8	27.5	8.3	22.4	8.1	14.3	11.6
Instruments & controls	58.3	0.8	27.3	2.6	36.2	10.0	38.8	10.4	13.8	11.7
Consumer goods	20.8	0.1	15.0	3.1	6.6	9.2	12.1	11.1	24.1	7.3

[a] For each country and technical field, 'National' + 'Foreign' + 'Other (not shown)' = 100.

cal activities. It shows that the predominance of foreign firms is even more pronounced in electrical and electronics technologies, where foreign firms are more than twice as important as national firms in British technological activities. Since the late 1960s, the proportion of innovative activities in electronics accounted for by UK owned companies has declined rapidly, so that in the mid-1990s around 50 per cent or more of the total was performed by foreign firms in four out of the five electronics fields (computers, semiconductors, image & sound and photography & photocopying). In three out of these four fields, the UK's largest electronics company, GEC, has seen its share decline to around half its value in 1969–73. No other major OECD country has a similar pattern.

2.5 Internationalisation in leading-edge companies: home country vs host country advantages

The above type of analysis has been criticised for concentrating on averages, and thereby missing important changes that may be taking place amongst those leading-edge companies with the most advanced policies and practices in the internationalisation of their technological activities. In particular, it is argued that the emphasis in internationalised technological activities is shifting from adaptation of products and processes to local market conditions, to the monitoring of host-country strengths in technological developments, and to the full-scale development and commercialisation of products and processes from foreign locations.

For this reason, we have recently undertaken an analysis of the 220 companies with the highest volume of patenting activities outside their home countries (Patel and Vega, 1998), in order to assess the extent to which companies locate their technological activities in a foreign location purely on the basis of the technological advantage of that location or on the basis of such advantage created at home. The main conclusions of this analysis are the following.

- In more than 75 per cent of cases, firms locate their technology abroad in core fields where they are strong at home. In 10 per cent of cases, firms establish technological activities abroad in fields of domestic weakness, in order to exploit the technological advantage of the host country.
- The largest increases in foreign technological activities, especially for chemical and pharmaceutical companies, have been in technical fields where there are complementary strengths between the domestic activity of a company and the host country.

These results suggest that adapting products and processes and materials to suit foreign markets and providing technical support to offshore manufacturing plants remain major factors in the establishment of foreign R&D activities. They are also consistent with the notion that firms are increasingly engaging in small scale activities to monitor and scan new technological developments in centres of excellence in foreign countries within their areas of existing strength. However there is very little evidence to suggest that most firms primarily go abroad to compensate for their core weaknesses at home.

3 Underlying causes

These gradual rather than spectacular trends in the nature and extent of the internationalisation – not globalisation – of corporate innovative activities cannot be interpreted as a major discontinuity or 'paradigm shift'. Indeed, Cantwell (1995) has shown that for some firms (mainly those based in the US and Sweden) the degree of internationalisation of their innovative activities was higher in the 1930s than in the 1970s and the early 1980s. Instead, we shall interpret the observed changes as the consequences of four continuous trends since the foundations of the present international economic system were laid after the Second World War. Taken together, they have placed nationally based linkages between the publicly funded science base and locally owned corporate practitioners under growing strain. The trends are:

- liberalisation of international exchanges
- uneven rates of national technological development
- increasing pressures of competition
- increasing range of fields of potentially useful technological knowledge.

3.1 1950s & 1960s: USA dominant

Table 9.5 attempts to summarise the effects of these changes. After the Second World War, the US was the technologically dominant power, and this was reflected in the reality of international economic and technological exchanges. The US trade advantage compared with the rest of the world was in technology-intensive products, and trade liberalisation ensured that US technology was diffused internationally through exports of capital and intermediate goods. The subsequent liberalisation of capital movements widened the channels of international diffusion to include licensing and direct foreign investment, particularly for products whose technology had stabilised, thereby enabling US firms to profit from the lower costs of foreign production. Outside the military and health spheres, the US government's role in technical change was restricted mainly to funding basic research in universities in what was seen as a necessary underpinning of the US system of innovation.

The underlying analytical structures that influenced policy reflected these realities. Vernon (1966) explained US comparative trade advantage by innovative leads resulting from labour-saving innovations induced by relatively high wages, and US foreign investment as 'trickle-down' of the production of technically mature products in lower wage regions. The international diffusion of technology was embodied in exports of capital and intermediate goods, and in transfers of production skills. Nelson (1959) and Arrow (1962) employed a one-country model in explaining the major role of the US government in funding of basic research. They argued that, since the output of basic research is codified information that is costly to produce but virtually costless to transfer and re-use, private funding and appropriation of the results of basic research will restrict the use of basic research below the social optimum: hence public funding of published research is justified on grounds of economic efficiency.

Table 9.5 *Changing modes & models of world innovation & diffusion*

Period & conditions	Govt policy for S&T in OECD countries	Corporate policy for locating R&D	International technological exchanges
1950s & 1960s • USA dominant • Freer trade • Growth of US foreign direct investment (FDI)	• Fund basic research to provide applicable info. for corporate practitioners	• At home. Incorporate innovations in processes and products, esp. exports. • Choice between exports, licensing & FDI, depending on product (*not* industry) maturity	• Embodied in US capital goods • Disembodied skills to improve and adapt
1970s • Spread of national capacities to imitate & innovate	• Protect basic research • Protect inefficient champions	• Adaptive R&D linked to FDI, licensing, & reverse engineering	• Disembodied skills to improve and adapt
1980s & 1990s • Several centres of world class innovation	• Stress local advantages of basic research – skills, networks • Basic research to attract world best practice firms • Stress support for *diffusion* (best practice foreign firms) rather than for *innovation* (potentially inefficient national champions)	• Access world centres of technological excellence through foreign labs & networks	• Multilateral exchanges of world class technological skills
• Growth in range of applicable technologies		• Increase range of required corporate competencies	

3.2 1970s: *new national capacities to imitate and innovate*

By the 1970s, the realities of uneven rates of national technological development were beginning to leave these analytical structures behind. It became clear from experience that technical change in catching-up countries required more than the transfer of production equipment and production skills. It also required a deliberate investment in improvement-generating activities like quality control, training, production engineering and R&D: hence the growth of adaptive R&D activities associated with direct foreign investment and other forms of technology acquisition (Bell and Pavitt, 1993).

In addition, new national centres of world-level innovation had become established, and began to export technology-intensive goods. Germany and other countries of continental north-west Europe re-established the positions of eminence that they had held before the Second World War. Japan – and later in the 1980s South Korea and Taiwan – emerged as new members of the still exclusive club of world-level innovators.

3.3 *Adjusting public policies: from protection to inward diffusion*

The emergence of these competitors in innovation posed difficult and novel policy problems for the countries that had perceived themselves as world leaders in science and technology – the US and (to a much lesser extent) the UK. Increasing liberalisation meant that international weaknesses in corporate innovativeness and competitiveness were felt more quickly. The immediate reactions tended to be a protective techno-nationalism to try to keep foreigners away from both domestic basic research and flagging national corporate champions. The same analytical framework that justified public subsidy for basic research in a one-country framework became awkward in a multi-country framework. If the ease of transfer and re-use of the results of basic research justified a public subsidy, it also meant that the knowledge could flow freely and easily in or out of the country. Hence, accusations against foreign competitors – especially Japan – for being 'free-riders' not contributing to 'the world pool of knowledge' alternated with recommendations to cut back on basic research and to import basic knowledge like the other countries were (apparently) doing. This unedifying interlude came to a close with the growing recognition[13] that the main economic benefits of basic research are not codified knowledge, but – as we have already anticipated in the introduction to this chapter – research skills, problem-solving techniques, and membership of international research networks. In this context, knowledge and its exploitation do not come free or cheap, and Japan and neighbouring countries have invested amply themselves in research training and in establishing international networks.[14]

Similarly, the protection of uncompetitive national corporate champions on grounds that included their privileged links with the national science base have turned out to be expensive. Governments in most cases have not been effective through direct intervention in transforming inefficient into efficient enterprises.[15] Policies have shifted from the promotion of innovation in nationally controlled firms to the promotion of the inward diffusion of world best practice. As has been pointed out elsewhere, R&D activities are essential inputs to both diffusion (catching up) and innovation (forging ahead).[16] To be effective, R&D for diffusion must be closely coupled to foreign sources of knowledge and technique through such commercial channels as licensing, purchases of capital goods, inward direct investment, and participation in international supply chains (such as OEM agreements).[17] R&D related to innovation must in addition forge close and flexible links with the latest advances in public knowledge and techniques emerging from the basic research in local universities.

These somewhat differing inputs required for innovation and diffusion were recognised back in the 1960s by Vernon (1960, 1966) in his analysis of causes of business agglomeration in New York City in the late 1950s, and later reflected in his classic and influential paper on the product cycle and international trade. The difference between the first stage of the product cycle (innovation) and the second stage (diffusion) is that, in the first stage, the product and process parameters are not fully understood and fixed, so that flexible and personal links are required, both within the firm between its functional departments, and outside the firm with potential providers of knowledge, complementary inputs, and so on. In the second and later stages, parameters are better understood and defined, so that the international transfer of knowledge related to production becomes easier.[18] In other words, national systems of innovation are in fact made up of two, closely related elements: national systems of diffusion and national systems of innovation, narrowly defined. Over the past thirty years, national systems of diffusion have been reinforced in their effectiveness by the growing liberalisation of international trade and investment. At the same time, the consequent increase in competition, coupled with the progressive spread of technological activities to a wider range of countries, has put national systems of innovation under strain.

Policies favouring inward diffusion at the expense of nationally based innovation were initially pursued most vigorously by the UK Thatcher governments of the 1980s in the automobile industry, where all large-scale producers are now foreign owned.[19] Similarly, in the electronics industry, government protection for defence and telecommunications equipment was reduced and foreign companies encouraged to enter the UK market. The consequences have been a loosening of the links between the national science base on the one hand, and both large UK owned firms and the UK production activities of foreign owned firms on the other. Instead, links have strengthened between the UK science base, and both small UK based firms and foreign firms with R&D activities in the UK.[20] These small, R&D performing UK based firms have specialised in international niche markets like application software, automobile design consultancy and Formula One racing cars, very often feeding into the R&D activities of foreign firms in foreign countries.

3.4   Adjusting corporate policies: in search of global excellence

The advantages to business firms of physical agglomeration of R&D activities, and of close linkages to the national science base, are overwhelming for the launching of major innovations. Yet, as we have seen in section 2.5 above, some large firms are increasingly locating some of their foreign R&D activities in fields of local technological strength rather than as a support for local production in fields of technological weakness.[21] This tends to happen when firms cannot create all the conditions necessary for launching major innovations in their home base. In part, this reflects the increasing range of potentially useful technologies that large firms need to master (Granstrand, Hakanson and Sjolander, 1992), so that the public science base cannot guarantee to provide the required skill and knowledge in all the important emergent fields with equal effectiveness – a problem that first affects large firms based in small countries. In part, it reflects institutional rigidities within the public science base.

In Japan, for example, the national science base cannot at present meet satisfactorily all the requirements of Japanese high technology firms. In spite of the rapid growth of Japanese scientific output and increased funding, institutional reforms (like the introduction of project and programme funding judged by peers) are required in Japanese universities, in order to increase creativity. In the meantime, Japanese firms have been developing world networks, involving the establishment of small but high quality laboratories near world centres of excellence. Similarly, German firms in the chemical industry established strong links with major US research in molecular biology and related medical fields, due to the backwardness of the national science base in this field.

In both these cases, the long-term expectation is that the knowledge and skill learned in foreign countries will be transferred back to the home country, mainly embodied in trained researchers. In the meantime, reforms of the science base in the home country will improve its usefulness to local firms. Japanese science is increasing its overall share of the world's papers and citations (Lattimore and Revesz, 1996). German public research in biotechnology has improved considerably (Sharp and Momma, 1997). It is expected in both cases that a system of innovation will be restored that links the national science base with national firms.

Such policies often are not feasible for large firms based in small countries, which often can find neither domestic sources of the specific skills they require, nor advanced users to test their major innovations. As a consequence they establish foreign R&D not for purposes of local market adaptation, nor for monitoring and networks, but to develop families of products in a specific field for sale in world markets. The location of such activities is heavily influenced by the local availability of the required skills and knowledge.[22] This explains the high share of foreign innovative activities of large firms based in the Netherlands, Sweden and Switzerland, as well as the high proportion located in regions with similar technological strengths. Whilst this international spread of corporate R&D activities overcomes the constraints of size of the national system of innovation, language and mobility barriers decrease corporate flexibility to re-deploy and re-combine person-embodied skills and resources in the light of changing technological opportunities.

4 Conclusions

A national system of innovation in which a strong national science base is coupled to innovative, competitive and often large national firms has many attractions. For the public policymaker, it maximises national returns for public and private investments in R&D in terms of efficiency and competitiveness. For the corporate manager, it reduces the considerable difficulties of managing the development and commercialisation of major innovations across national boundaries. This is enough to explain both the observed persistence of the strong links between national scientific and national technological performance, and the observed low level of internationalisation of innovative activities compared with other corporate functions. However, national systems of innovation thus defined are coming under increasing strain, because a combination of factors – liberalisation, uneven rates of national technological development, and the increasing range of technological skills that firms must master – are creating imbalances or mis-matches between the publicly funded science base and the corporate champions that they have served. Two broad categories can be identified.

In the first, uncompetitive national firms cannot sustain their demands on the public science base, which comes to be linked increasingly to the demands of foreign firms, either through their locally established R&D and related activities, or through their requirements for high technology components, sub-systems and services. This has happened in the UK automobile and electronics sectors over the past twenty years. It is an outcome that is less preferable than strong British firms linking effectively British automobile design and solid state physics skills to product developments for mass markets. But this was not feasible with British based firms in the past, and will be feasible in the foreseeable future only with foreign controlled firms. In the meantime, the research of Mason and Wagner (1999) shows that the present decoupled and messy British innovation system in the electronics industry is economically more rewarding than the more tidy but inefficient one of the past.

In the second category, national public science systems are not always able to provide the skills and knowledge that national innovating firms require. In some instances – as we saw for Germany and Japan – the internationalisation of corporate R&D will be seen as part of the remedy to improve the national science base. In others – especially firms from smaller countries – substantial internationalisation of R&D will be a permanent part of the solution, even if it does reduce the flexibility to re-combine resources in the light of new opportunities.

Finally, we must point out that the ad hoc nature of our analysis reflects the weaknesses of the analytical structures on which they might be based. A theory of public expenditure on the science base that assumes just one country, and that knowledge is information, no longer helps very much. Nor do the recent insights of the new growth theory, which stress the differences in knowledge externalities – or spillovers – in explaining differences in national rates of growth. The practical implications of this approach are that governments should fund activities with lots of externalities – like academic research and related training. But historical experience gives scant support to such a policy. Britain's substantial lead over Germany and Japan in the 1960s in academic research did not lead to faster rates of technical change and manufacturing growth. On the contrary, Germany and Japan caught up with and overtook Britain in technology; and the former now has done so in science and the latter is catching up. This experience if anything supports the reverse causality – first proposed by de Tocqueville and Marx – that dynamic capitalists generate more local demands for knowledge and skills than do undynamic ones. The problem is that – on the basis of the past record – most governments do not know how to transform undynamic capitalist firms into dynamic ones. Certainly, the US experience shows that massive government subsidies for defence and health R&D did create positive externalities in the form of major commercial technological opportunities in semiconductors and biotechnology. But these owed less to the foresight of the US government and to its acquaintance with the ideas of the new growth theorists, than to the US public's fear of communism and cancer.

In these circumstances, we can offer only two tentative conclusions about the future.[23] The first is that national systems of innovation will increasingly involve linkages between the local science base and foreign firms. The degree to which these links will grow beyond monitoring and training activities into local product development and commercialisation remains unclear. Second, the relative growth of international corporate linkages to foreign centres of scientific and technological excellence – as compared to those related to local adaptation and 'trickle down' – may help to explain the increasingly specialised patterns of national technological activities (Archibugi and Pianta, 1992). By themselves, foreign corporations are unlikely to create new world-class centres of technological activity.

Notes

1. This chapter is based on research funding from the EC-TSER programme (SOE1-CT95-1006) of the European Union. We thank John Cantwell for his comments on an earlier draft. The responsibility for the final draft is our own.
2. On the former, see Prais (1993).
3. The currently fashionable term is "knowledge production".
4. For the importance of distance, see Jaffe (1989). For the usefulness of academic research, see Martin and Salter (1996)
5. Compare, for example, Dunning (1992), who argues that the links between globalised firms and their technological home base are becoming weaker, Porter (1990) who argues that they continue to be strong, and Vernon (1979) who takes a position in between.
6. See, for example, Cantwell (1992) and Patel (1995, 1996).
7. See also Patel (1995, 1996).
8. In terms of nationality, 136 are European, 128 American and 95 Japanese.
9. The larger share of foreign patenting than foreign R&D for the European and Japanese based firms probably reflects differing populations, since our patenting data cover only large firms which typically have a higher share of foreign activities. The lower share of foreign patenting for the US based firms probably reflects the higher propensity of US based firms to patent in the US from their domestic

base than from foreign countries.
10 Country of the headquarters of the parent group.
11 The patenting data underestimate the foreign share of technological activities in the US, because of the lower propensity of foreign firms to patent in the US than US-based firms
12 See also Cantwell (1995).
13 By policymakers and academics. Business practitioners have known better for a long time.
14 Japanese output of basic research has also been increasing rapidly, in terms both of numbers of refereed papers published and the growing frequency with which they are cited. The same is true for South Korea and Taiwan. See Lattimore and Revesz (1996).
15 There were exceptions, such as telecommunications in France in the 1970s.
16 See Cohen and Levinthal, 1989.
17 See Hobday, 1995, Arundel, van de Paal and Soete, 1995.
18 In the context of this chapter, we are applying the product cycle concept to a product *model* (e.g. VW Golf), and not to a product *class* (e.g. automobiles). For the former, the product cycle concept is empirically well grounded. For the latter, its validity can and is questioned. Lack of clarity about the unit of analysis for the product cycle continues to cause enormous confusion.
19 Some other European countries were slow to follow similar policies, even accusing the UK of being a Trojan horse allowing US and Japanese competitors easy access to the European market. But even the French government recently welcomed a large investment by a Japanese automobile company.
20 The major corporate contributions to basic research in solid state physics at the Cavendish Laboratory in Cambridge come from foreign owned companies.
21 See also Gerybadze and Reger, 1997.
22 Thus, one of the examples cited by Solvell (1997) is Nestlé, which has located its home base for pasta in Italy, and for confectionery in Great Britain.
23 For similar reasoning, see Cantwell (1998).

References

Archibugi, D. and Pianta, M. (1992), *The Technological Specialisation of Advanced Countries*, report to the EC on International Science and Technology Activities, Dordrecht, Kluwer.

Arrow, K. (1962), 'Economic welfare and the allocation of resources for invention', in Nelson R. (ed.), *The Rate and Direction of Inventive Activity*, Princeton, NJ, Princeton University Press.

Arundel, A., van de Paal, G. and Soete, L. (1995), *Innovation Strategies of Europe's Largest Industrial Firms* (PACE Report), MERIT, University of Limbourg, Maastricht.

Bell, M. and Pavitt, K. (1993), 'Technological accumulation and industrial growth:

contrasts between developed and developing countries', *Industrial and Corporate Change*, 2, pp. 157–210.

Cantwell, J. (1992), 'The internationalisation of technological activity and its implications of competitiveness', in Granstrand *et al.* (1992).

—— (1995), 'The globalisation of technology: what remains of the product cycle model?', *Cambridge Journal of Economics*, 19, pp. 155–74.

—— (1998), 'Innovation as the principal source of growth in the global economy', in Archibugi, D., Howells, G. and Michie, J. (eds.), *Innovation in the Global Economy*, Cambridge, Cambridge University Press.

Cohen, W. and Levinthal, D. (1989), 'Innovation and learning: the two faces of R&D', *Economic Journal*, 99, pp. 569–96.

Dunning, J. (1992), 'Multinational enterprise and the globalization of innovatory capacity' in Granstrand *et al.* (1992).

EC (1997), *Technology Policy in the Context of Internationalisation of R&D and Innovation: How to Strengthen Europe's Competitive Advantage in Technology*, Draft Report for the ETAN Pilot Activity, DG XII.

Edquist, C. (1997) (ed.), *Systems of Innovation: Technology Institutions and Organisations*, London, Pinter.

Etemad, H. and Seguin-Dulude, L. (1987), 'Patenting patterns in 25 multinational enterprises', *Technovation*, 7, 1, pp. 1–15.

Fagerberg, J. (1994), 'Technology and international differences in growth rates', *Journal of Economic Literature*, 32, pp. 1147–75.

Freeman, C. (1995), 'The "National Systems of Innovation" in historical perspective', *Cambridge Journal of Economics*, 19 (1), pp. 5–24.

Gerybadze, A. and Reger, G. (1997), 'Globalization of R&D: recent changes in the management of innovation in transnational corporations', Discussion Paper on International Management and Innovation, Stuttgart.

Granstrand, O., Hakanson, L. and Sjolander, S. (1992) (eds), *Technology Management and International Business: Internationalisation of R&D and Technology*, Chichester, Wiley.

Hicks, D., Izard, P. and Martin, B. (1996), 'A morphology of Japanese and European corporate networks', *Research Policy*, 25, pp. 359–78.

Hobday, M. (1995), *Innovation in East Asia*, Guildford, Elgar.

Jaffe, A. (1989), 'Real effects of academic research', *American Economic Review*, 79, pp. 957–70.

Kealey, T. (1996), *The Economic Laws of Scientific Research*, London, Macmillan.

Lattimore, R. and Revesz, J. (1996), *Australian Science: Performance from Published Papers*, Bureau of Industry Economics, Report 96/3, Canberra, Australian Government Printing Office.

Lundvall, B-Å. (1992) (ed.), *National Systems of Innovation: Towards a Theory of Innovation and Interactive Learning*, London, Pinter.

Martin, B. and Salter, A. (1996), *The Relationship between Publicly funded Basic Research and Economic Performance: a SPRU Review*, Report for HM Treasury, Science Policy Research Unit, University of Sussex.

Mason, G. and Wagner, K. (1999), 'Knowledge transfer and innovation in Germany and Britain: "Intermediate Institution" models of knowledge transfer under strain?', *Industry and Innovation*, 6, 1, pp. 85–109.

Narin, F., Hamilton, K. and Olivastro, D. (1997), 'The increasing linkage between US technology and public science', *Research Policy*, 26, pp. 317–30.

Nelson, R. (1959), 'The simple economics of basic scientific research', *Journal of Political Economy*, 67, pp. 297–306.

— (1993)(ed.), *National Innovation Systems: a Comparative Analysis*, Oxford, Oxford University Press.

OECD (1997), *Internationalisation of Industrial R&D: Patterns and Trends*, DSTI/IND/STP/SWP/NESTI(97)2, Paris, October.

Patel, P. (1995), 'The localised production of global technology', *Cambridge Journal of Economics*, 19, pp. 141–53.

— (1996), 'Are large firms internationalizing the generation of technology? Some new evidence', *IEEE Transactions on Engineering Management*, 43, pp. 41–7.

Patel, P. and Pavitt, K. (1991), 'Large firms in the production of the world's technology: an important case of non-globalisation', *Journal of International Business Studies*, 22, pp. 1–21.

— (1994), 'National innovation systems: why they are important, and how they might be measured and compared', *Economics of Innovation and New Technology*, 3, pp. 77–95.

Patel, P. and Vega, M. (1998), 'Technology strategies of large European firms'. Draft Final Report for 'Strategic Analysis for European S&T Policy Intelligence' Project, funded by EC Targeted Socio-Economic Research Programme, Brighton, February.

— (1999), 'Patterns of internationalisation of corporate technology: location versus home country advantages', *Research Policy*.

Porter, M. (1990), *The Competitive Advantage of Nations*, London, Macmillan.

Prais, S. (1993), 'Economic performance and education: the nature of Britain's deficiencies', Discussion Paper No. 52, National Institute of Economic and Social Research, London.

Rothwell, R. (1977), 'The characteristics of successful innovators and technically progressive firms', *R&D Management*, 7, pp. 191–206.

Sharp, M and Momma, S. (1997), 'New biotechnology firms in Germany', Paper presented at the Fifth Annual High Technology Small Firms Conference, Manchester 28 May, Science Policy Research Unit, University of Sussex.

Solvell, O. (1997), 'Local innovation in global markets – implications for European science and technology policy', RP 97/14, Institute for International Business, Stockholm School of Economics.

Vernon, R. (1960) *Metropolis, 1985*, Cambridge, Mass., Harvard University Press.

— (1966), 'International investment and international trade in the product cycle', *Quarterly Journal of Economics*, 80, 2, pp. 190–207.

— (1979), 'The product cycle hypothesis in a new international environment', *Oxford Bulletin of Economics and Statistics*, 41, pp. 255–67.

10 Agglomeration economies, technology spillovers and company productivity growth

PAUL GEROSKI, IAN SMALL and CHRISTOPHER WALTERS[1]

1 Introduction

Most people believe that the natural starting point for any examination of the determinants of superior company performance lies inside the firm, in how effectively it deploys its knowledge, skills and core capabilities. However, firms sometimes benefit from externalities created by the superior performance of their rivals. For example, external economies of scale created by sharing a common indivisible input or a deep and well functioning local labour market are often used to account for the superior performance of firms located in industrial clusters like Silicon Valley or Route 128. Similarly, the ability to absorb other firms' technological spillovers is an important feature of the outstanding performance of some firms, particularly in high technology sectors. The existence of either type of externality is interesting and important because it breaks the close link which ought to exist between the actions a firm takes and the performance it realises. Further, such externalities may generate macroeconomic consequences which sometimes seem to be more than the sum of their parts.

It is worth distinguishing two types of external effects: *agglomeration economies*, which directly induce a positive correlation in productivity growth across firms, and *technology spillovers*, where the correlation between productivity growth rates is driven by a correlation between the incidence in innovative activity between firms. Previous work on agglomeration economies has typically used industry level data. Modest agglomeration effects have been observed, but the simple fact is that aggregation makes agglomeration effects difficult to distinguish from internal economies of scale. This, of course, means that industry level work is likely to *understate* the size of agglomeration economies. The evidence on technological spillovers from past work is rather more plentiful, involving both firm level econometric studies and survey/case study evidence. However, this work commonly uses measures of R&D inputs to help identify spillovers rather than measures of innovative output. Output measures identify the development of new products and processes more directly than measures of expenditure on research and development do, and, as a consequence, reflect differences between firms in their ability to extract innovative output from a given R&D spend. If the productivity of R&D (expressed in terms of innovative outputs) is correlated with other measures of corporate productivity, then using R&D to measure spillovers may create an error of measurement which leads to an *understatement* in the size of the effects of innovation and innovation spillovers on productivity growth.

In what follows, we report work using techniques developed in the literature to assess the importance of agglomeration economies and technology spillovers using data drawn from a sample of 216 large UK firms observed over the period 1974–90. Our expectations are that exploring these issues at a more microeconomic level of aggregation than is common will yield estimates of external effects which are somewhat larger than those reported in more aggregated studies. The interesting question, of course, is whether they are large enough to account for a substantial fraction of the variation in company productivity growth performance which we observe. The answer to this question turns out to be 'no'. Although we observe correlations which are consistent with the existence of external economies, they are modest in size and do not really increase our ability to explain variations in the productivity growth performance of firms over time. The plan of the chapter is as follows. In section 2, we introduce our data and review the basic productivity growth framework that we will be using. Section 3 describes our work trying to identify both types of externality, while section 4 reports a number of further experiments designed to check the robustness of these results. Our conclusions are contained in section 5.

2 The sample and data

Our baseline sample consists of an unbalanced panel of 216 large UK firms observed over the period 1974–90. It was drawn from a frame consisting of 4,082 firms taken from the Exstat database, and it includes all firms with at least eight continuous observations and for whom capital stock data could be constructed (see Appendix I to this chapter for details).[2] The average size of the firms in the sample was 3055 employees in 1974 (with a standard deviation of 4327 and a range of [157, 29076]); by 1990, the average size of the sample had risen to 3637 employees.

We start by examining the Solow residual, defined as:

$$\theta_i(t) \equiv \Delta y_i(t) - \alpha \Delta l_i(t) - \beta \Delta k_i(t) \tag{1}$$

where $\Delta y_i(t)$ is the first difference in the log of value added for firm i at time t, $\Delta l_i(t)$ is the first difference in the log of employment and $\Delta k_i(t)$ is the first difference in the log of the real capital stock. Following standard practice, we assume that $\beta = (1-\alpha)$; and hence that constant returns to scale prevail. This enables us to write

$$\Delta y_i(t) = \gamma \Delta x_i(t) + \theta_i(t) \tag{2}$$

where $\Delta x_i(t) \equiv \alpha \Delta l_i(t) + (1-\alpha)\Delta k_i(t)$. For the remainder of this section, we will assume that $\gamma = 1$, an assumption that will be unwarranted if the production function displays non-constant returns to scale or if firm i has enough market power to elevate its price above marginal costs. When $\gamma > 1$, variations in inputs over the cycle induce greater variations in output, resulting in a pro-cyclical pattern of variation in $\theta_i(t)$.[3]

Three properties of $\theta_i(t)$ calculated under these assumptions are worth noting. First, the data display mainly variation for firms over time (within variation accounts for nearly 92 per cent of the total variation) and, what is more, this within variation is mostly idiosyncratic. Rankings of firms by productivity growth rates are very unstable over time (spectacularly so in some cases), and partial correlations between cross-sections of firms ranked by productivity growth rates fall to zero very rapidly as the time elapsed between the cross sections increases.

Second, value added growth rates are more highly correlated with the rate of growth of labour than with the rate of growth of capital (the correlations are 0.3841 and 0.2595 respectively). Moreover, the cross-section correlation between the growth of value added and the two input growth rates increases markedly over time, rising from 0.1153 (0.3067) in 1974 to 0.3687 (0.6632) in 1990 for capital (labour). The cross-section correlation between the rate of growth of value added and the revenue share weighted sum of the two input growth rates, $\Delta x_i(t)$, ranges from 0.3062 in 1974 to 0.6704 in 1990. These correlations do not, however, rise secularly through the period. Rather, they vary (roughly) pro-cyclically, rising to relatively high levels in 1976–7 and from 1982 until the late 1980s, but starting low in the early mid-1970s and falling markedly in the recession centred on 1981. It is not entirely clear what lies behind this pattern of correlations, although strong pro-cyclical and weak counter-cyclical correlations between output growth and input growth may generate what appears to be pro-cyclical productivity growth rates over the period as a whole.

Third, cyclical influences affect not only mean productivity growth rates, but also their variability over time. Mean annual total factor productivity growth over the whole seventeen-year period is 0.01, with a standard deviation of 0.2579. More interestingly, during the recession centred on 1981, mean productivity growth rates are unusually low (as one expects) while, at the same time, the cross-section variation in productivity growth rates across firms is unusually high. Further, over the sample as a whole, the mean and the variance are negatively correlated, an observation which is true even if the few years centring on 1981 are ignored. This suggests that the rate of productivity growth of some firms is more cyclically sensitive than that of others. Amongst other things, the observation is consistent with recent work suggesting that the effects of recession on firms are very selective, with a small number of firms being very badly affected.[4] It is also not obviously consistent with the assertion that strong agglomeration effects exist.

3 Agglomeration economies and technology spillovers

Our goal in what follows is to measure the size of agglomeration economies and technology spillovers between firms. Both types of externality are difficult to observe directly. Occasionally, one can observe a particular action taken by one firm which benefits a number of other firms, and flows of information can sometimes be observed through patent citations or by case study methods. However, it is difficult to imagine how one might directly observe or comprehensively measure intangibles like external effects for a large sample of firms whose performance is observed over a long period of time. As a consequence, one is forced to try to make inferences about the existence of externalities by looking at their consequences. Agglomeration economies ought to directly induce a positive correlation between the productivity growth (or other measures of performance) of otherwise unrelated firms, while technological spillovers ought to induce a correlation between the innovative output of one firm and either the innovative output or the productivity performance of other firms.

3.1 Agglomeration economies

Agglomeration economies arise whenever the simultaneous expansion of firms up and down the value chain or across particular markets facilitates transactions, stimulates the emergence of common inputs or creates rich inter-firm flows of information. Thus, the mutual interaction between a number of sectors which simultaneously industrialise, or a group of firms which simultaneously implement a new process or operating system, can deepen and enrich markets to such a degree that projects which are unprofitable if undertaken singly may become profitable if undertaken simultaneously. Similarly, the introduction of 'general purpose technologies', rail or road networks, co-operative marketing agencies or the creation of 'thick' markets for particular inputs provide a platform on which a number of firms can simultaneously successfully expand.[5] External economies in the form of agglomeration effects are often easy to observe in specific regions or locations (e.g. Silicon Valley), and, indeed, they sometimes help to encourage an increase in the localisation of particular types of economic activity.[6]

We need to embody these ideas in a simple empirical vehicle. If firm i's productivity growth depends on the activity of firms in the same industry as itself or in the economy as a whole, then the size of agglomeration economies should be reflected in the strength of the correlation between i's productivity growth and increases in industry or macroeconomic activity. Thus, if $\Delta y(t) \equiv \Sigma \Delta y_i(t)$, then including the term $\phi \Delta y(t)$ into (2) allows the growth of all other firms in the economy to affect that of firm i. In particular, if

$$\Delta y_i(t) = \gamma \Delta x_i(t) + \phi \Delta y(t) + \theta_i(t) \tag{3}$$

then $\phi > 0$ is consistent with the hypothesis that agglomeration economies are present. Notice that the measured size of agglomeration economies can be sensitive to the degree of aggregation at which they are measured. Aggregating (3) over all firms i yields

$$\Delta y(t) = \delta \Delta x(t) + \Theta(t) \tag{4}$$

where $\delta \equiv \gamma/(1-\phi)$ and $\Theta(t)$ is the weighted sum of the $\theta_i(t)$ divided by $(1-\phi)$. Since $0 < \phi < 1$ almost surely, it follows that estimates of (internal) economies of scale at the aggregate level (i.e. estimated values of δ) will

Table 10.1 *Estimates of equation (5) at company level, two-digit level and macro level*[a]

Independent vars.	(i) (company)	(ii) (two-digit)	(iii) (macro)	(iv)	(v)
$\Delta x_i(t)$	0.8054 (15.86)	0.8122 (4.89)	1.4464 (2.57)	0.7887 (14.48)	0.7887 (14.44)
$\Delta y_I(t)$	0.4568 (4.50)	–	–	0.4320 (4.08)	0.4407 (4.17)
$\Delta y(t)$	0.4664 (4.70)	0.4958 (2.87)	–	0.5066 (4.74)	0.4977 (4.63)
$\Delta y_i(t-1)$	–0.1920 (5.63)	–0.1734 (2.00)	–0.1206 (–0.58)	–0.2151 (5.96)	–0.2156 (5.95)
CNST	0.0265 (5.03)	0.1777 (1.28)	–0.0116 (0.44)	0.0351 (6.30)	0.0379 (5.17)
$\overline{R}^2$	0.2207	0.1945	0.1784	0.2160	0.2145
L.L'h	–42.398	106.372	21.319	–57.28	–56.31
Standard error of regression	0.2464	0.1509	0.7608	0.2486	0.2488

Note:
[a]Dependent variable $\Delta y_i(t)$ is the rate of growth of firm i's value added. Absolute values of t-statistics are given in brackets below the estimated coefficients, and all estimates are heteroscedastic consistent. $\Delta x_i(t)$ is the revenue share weighted average of the rate of growth of labour and capital, $\Delta y_I(t)$ is the rate of growth of net (of firm i's value added) two digit value added and $\Delta y(t)$ is the rate of growth of net (of firm i's value added) aggregate value added. Regression (i) uses company data, regression (ii) aggregates this up to two-digit levels, and regression (iii) aggregates all firms into a single 'representative firm'. Regressions (i)–(iii) use a sample of 216 firms over the period 1974–90. Finally, regression (iv) replicates (i) for the sample over the period of 1974–88, while (v) extends (iv) to include the following technology spillover terms: $0.0013*P_i(t) - 0.0013*P_i(t-1) - 0.0054*P_i(t-2) + 0.0047*P_i(t-3) + 0.0036*P_i(t-4) - 0.0040* P_i(t)$. All of the individual t-statistics on these terms are less than unity in absolute value.

almost surely overstate (internal) returns to scale at the micro level (measured by γ) whenever agglomeration economies are present. Intuitively, the aggregation of seemingly interdependent units 'internalises' any agglomeration effects which arise between them, making the combined output growth of aggregated units more responsive to changes in inputs than is the case for any unit taken in isolation.[7]

Column (i) in table 10.1 shows an estimate of a slightly more general version of (3), namely

$$\Delta y_i(t) = \gamma \Delta x_i(t) + \phi_1 \Delta y(t) + \phi_2 \Delta y_I(t) + \varphi \Delta y_i(t-1) + \theta_i(t) \tag{5}$$

The two changes which transform (3) into (5) are the inclusion of $\Delta y_i(t-1)$ to pick up any lags in response associated with adjustment costs, and the inclusion of two terms to pick up agglomeration economies: $\Delta y_I(t)$ is the rate of growth of real value added in the industry I in which firm i operates defined *net* of i's real value added growth, while $\Delta y(t)$ is aggregate real value added growth defined *net* of that of industry I. This distinction is designed to allow for the entirely reasonable possibility that agglomeration effects may be stronger within industries than across the economy as a whole. By construction, (5) aggregates into (4) in much the same way as (3) does. Finally, we write

$$\theta_i(t) = \dot{\theta} + \mu_i(t) \tag{6}$$

where θ is a constant, and $\mu_i(t)$ is a white noise error process. Putting (6) into (5) produces a regression model,

$$\Delta y_i(t) = \gamma \Delta x_i(t) + \phi_1 \Delta y(t) + \phi_2 \Delta y_I(t) + \varphi \Delta y_i(t-1) + \theta + \mu_i(t) \tag{7}$$

The estimates of (7) are shown in regression (i). They suggest the existence of diminishing returns to scale at company level, something that seems to be a rather general feature of UK company data.[8] However, the coefficients on $\Delta y_I(t)$ and $\Delta y(t)$ are clearly statistically significant, and a 1 per cent increase in either two-digit or aggregate value added is associated with a 0.45 per cent increase in company productivity growth, *ceteris paribus*. Further, regressions (ii) and (iii) show that as we aggregate (7), returns to scale appear to be more substantial. The 'macro' equation estimates are 1.45 (or 1.30 in the long run) which could only be consistent with constant returns or diminishing returns if market power leads to a substantial mark-up of price over marginal costs. These results are in line with those reported earlier in the literature.[9] These estimates are also extremely robust to minor specification changes: we allowed γ to vary over time and across firms, entered labour and capital growth rates separately, used an error correction specification, included lagged values of $\Delta y_I(t)$ and $\Delta y(t)$, introduced fixed effects and re-estimated the model for several sub-periods of the data and for quartiles and quintiles of firms ranked by pre-sample average values of $\theta_i(t)$. If anything, these calculations suggested that the numbers on table 10.1 may slightly understate the size of agglomeration economies, returns to scale and agglomeration effects might have been higher in the 1980s than in the 1970s, and both very large and very small firms may have lower returns to scale than more moderately sized firms. Instrumenting input growth and the lagged dependent variable (see section 4 below) also produced only very modest effects on these estimates, and hence we would suggest that there is little evidence of simultaneity bias.

Our view is, however, that while agglomeration economies may be statistically 'significant', they are quantitatively unimportant. The fit on these regressions (and virtually all of those of this type that we have run) is pretty weak, meaning that most of the variations in productivity growth observed in the data are firm specific and idiosyncratic. Further, a comparison of (i) with and without the two agglomeration terms shows that they cause only a very small increase in explanatory power (R^2, for example, rises by only three percentage points) and produce no effects on the estimates of the other parameters. In addition, the coefficients on $\Delta y(t)$ and $\Delta y_I(t)$ are very similar in size, but most of the increase in the estimate of scale economies occurs when we aggregate up to a macro level equation (i.e. agglomeration economies do not appear to be particularly localised). This is hard to believe: the fact that external effects appear to be stronger across the economy as a whole rather than within industries seems a little implausible, and casts some doubt on whether we are actually capturing real agglomeration effects in these estimates.

As we noted above, company productivity growth rates are highly idiosyncratic. One way to get a handle on this is to compute the statistic

$$\Psi_i = \text{cov}(\theta_i, \Theta) / \text{var}(\theta_i) \tag{8}$$

which measures the degree to which firm i's productivity growth varies in step over time with that of other firms. $\Psi_i = 0$ if all of the variation in i's productivity growth is completely idiosyncratic, while $\Psi_i = 1$ if i's productivity growth is perfectly synchronised with aggregate productivity growth. The average value of Ψ_i across the 216 firms in our sample is 0.0421, while the range is [−0.030, 0.40]. In fact, just under a third of the calculated values of Ψ_i are negative, but the mean across positive values of Ψ_i is only 0.0863. All in all, these are fairly small numbers, and mean that less than 10 per cent of the variation in productivity growth rates across firms is common.[10]

In order to evaluate the importance of agglomeration effects, we re-ran regressions (i) and (ii) for sub-samples of firms ordered by two-digit industry, and found clear evidence that agglomeration effects associated with $\Delta y(t)$ were important only in Mechanical Engineering and Metal Manufacturing, while those associated with $\Delta y_I(t)$ were present only in Chemicals, Electrical Engineering, Textiles and Paper and Publishing. These are, on the whole, important innovation producing sectors (some are also regionally concentrated) and it is not, therefore, surprising that external economies appear to be relatively more important in these sectors than elsewhere. Thus, if agglomeration economies are important, it may be that they are so only in a few sectors, and work to the benefit of only a relatively few firms.

3.2 Technology spillovers

There is now quite a large literature of work reporting on the measurement of technology spillovers. It aims to uncover correlations between the innovative activity of some firm i, and that of other firms j. It is necessary to decide which firms j are likely to benefit from i's spillovers, and most scholars identify potential beneficiaries as those whose operations are 'similar to' or 'near' the operations of i. Typically, 'near' means in the same industry (horizontal relationships), or in an input–output chain (vertical relationships), or having a common scientific basis for whatever knowledge they share (or the same technological profile). Our data are not rich enough for us to do anything other than use horizontal classifications.[11] It is also useful to identify what it is that spills over. Most econometric studies use measures of i's R&D spend (or some measure of the stock of its R&D capital), but until recently the data have not been available at corporate level in the UK. Further, it is new ideas embodied in new products and processes that spill over, and measuring these by using information on innovative output is an attractive alternative to using R&D spending. We, therefore, use two measures of innovative output as a proxy for the knowledge which firm i might be generating and, therefore, spilling over to its rivals: the number of patents, P, and the number of 'major' innovations, IN, produced by firm i. This converts (7) into

$$\Delta y_i(t) = \gamma \Delta x_i(t) + \phi_1 \Delta y(t) + \phi_2 \Delta y_I(t) + \phi \Delta y_i(t-1)$$
$$+ \sum_j \{\eta_j^1 P_i(t-j) + \eta_j^2 IN_i(t-j)$$
$$+ \eta_j^3 P_I(t-j) + \eta_j^4 IN_I(t-j)\} + \theta + \mu_i(t) \qquad (9)$$

Unfortunately, our innovations data are only available until 1982 (patents are available until 1988), and this necessitates working with a slightly shorter sample. Regression (iv) on table 10.1 shows estimates of (7) (i.e.

it replicates (i)), while (v) shows the analogous estimates of (9) for the period 1974–88. The comparison of (iv) with (i) shows that our baseline estimates are not sensitive to this change in time period; comparing (iv) with (v) shows that the terms associated with patents individually and collectively make an insignificant contribution to the explanation of company productivity growth rates (F(6,2199) = 0.36 (0.904)). For the period 1974–82, estimates of returns to scale and agglomeration effects are a little higher ($\gamma = 0.94$, $\phi_1 = 0.84$, $\phi_2 = 0.57$), but all the own innovation terms and the innovation spillovers terms are individually insignificant, and can be collectively excluded without much effect (as can the innovations terms in these equations and in analogous equations using innovations only over a shorter period).[12]

These results may simply reflect the way that we have structured our calculations. Some firms are more likely to absorb spillovers than others located at the same distance from i, if only because a firm must often make a substantial investment in learning about a technology before it can learn anything from the work of its rivals. This means that firms who invest in R&D and innovate may be rather more likely to benefit from spillovers than others. To pursue this conjecture we re-ran (iv) on the sub-sample of firms that patented any time between 1974 and 1988 (too few firms produced a major innovation to make a split made on this basis very interesting). The most notable feature of the comparison between patenters and non-patenters is that the former were not affected by $\Delta y(t)$. They were, however, not more sensitive to the patenting and innovating activities of other firms than non-patenters were; i.e. they did not seem to benefit unduly from spillovers.[13]

We would conclude that innovative output has only very weak and imprecise effects on company productivity growth and, in addition, that technological spillovers between firms are generally fairly weak. This conclusion is not entirely consistent with the literature, which, for the most part, has used R&D measures of innovative activity to track spillovers and has also found positive effects of R&D on productivity. There are several potential explanations for this difference: the knowledge embodied in specific measures of innovative output may be much more specific than that reflected in general R&D spending, and may, therefore, spillover less or have less profound effects on productivity growth; users may capture most of the gains from innovative activity, leaving innovation producers with only modest apparent gains; horizontal classifications may be a poor way to track technology spillovers; and, last but not least, the multitude of factors which affect productivity growth rates and make it very idiosyncratically variable across firms and over time might mean that genuine but not overpoweringly strong effects on innovation or technological spillovers are just too hard to detect in the data, especially in a relatively short times series.[14]

One further explanation of these results is worth noting. A closer look at the basic data on the production of patents and innovations makes it plain why it is hard to detect any spillovers (or to measure them with any precision). Technology spillovers ought to induce a correlation between the innovative output of different firms, but the simple fact is that most firms are very irregular innovators (regardless of whether $IN_i(t)$ or $P_i(t)$ is used as a measure of innovation) and, more important, the timing of when they do innovate is not very highly related to that of other firms.[15] This is not a feature of data on R&D, which typically display a very high degree of between variation, and are fairly persistent over time. All of this says that corporate performance – at least as measured by productivity growth rates or the production of patents or innovations (but not by R&D spending or accounting profitability) – is very erratic. Periods of sustained superior performance are short and unpredictable, and are more or less independent of those of rival firms.

4 Robustness

The conclusions that we draw from analysing the baseline sample are that returns to scale are more modest at company level than at industry or, much more noticeably, at the macroeconomic level; that although the data are consistent with the existence of external effects, they are unlikely to account for more than a very small fraction of the total variation in company productivity growth rates; and that the effects of innovations, patents and technology spillovers on company productivity growth are so imprecisely estimated as to make it difficult to reject the null hypothesis that they have no effect.

There are a number of objections one might raise to these conclusions, and some of them are difficult to deal with using our current sample. We therefore put together three further samples: (1) an unbalanced panel of 327 firms observed over the period 1976–95; (2) an unbalanced panel of 491 firms, 1976–82; and (3) an unbalanced panel of 186 firms, 1983–95. Sample (1) contains the 216 firms in our baseline sample, but is much larger because we have calculated capital stocks in a rather different way. These samples are drawn from Datastream rather than Exstat, and Datastream changed the nature of the employment data it reported in 1982; further, after 1982, the profits data (used to calculate value added) gets a little better. Thus, splitting sample (1) at 1982 makes sense, and then recollecting the data to maximize the number of firms in the two resulting sub samples yields samples (2) and (3). Table 10.2 shows some of the results that we obtained by re-running regressions (i) from table 10.1 on these three additional samples. It has four panels: the first three display the results from running regression (i) on samples (1) to (3) respectively using real output as a dependent variable, while the fourth displays the results when real value added is used (on a slightly shortened sample).[16] The panels report eleven different regressions: fixed effects estimates, robust regression estimates, estimates imposing a single constant and then eight instrumental variables estimates of different types.[17]

Rather than reading through the table row by row, it seems sensible to try to group the results of this exercise around the issues that give rise to most concern. These are: the question of whether there are other explanations of pro-cyclicality which might be confounded with agglomeration effects, problems associated with potential endogeneity, sample biases and biases associated with various types of measurement error.

4.1 Other explanations of pro-cyclicality

The most salient feature of our data is the pro-cyclical variation in the mean of the Solow residual, and the search for external effects is, in effect, an exploration of one possible explanation of this phenomenon. If one confines attention to the macro level regression (iii) on table 10.1, then one can make a case for thinking that economies of scale created by agglomeration effects may account for the pro-cyclicality of the macro data on productivity growth. However, regression (i) makes it plain that economies of scale cannot be the explanation of why company productivity growth rates are positively correlated with movements in company real value added. This seems to be a very robust conclusion. The only regressions on table 10.2 which suggest anything other than apparent decreasing returns to scale at company level are some of the instrumental variables estimates, and in almost all of these cases, the instruments used in these estimates are not valid. Further, many of the estimates of ϕ_1, the coefficient on $\Delta y(t)$, and ϕ_2, the coefficient on $\Delta y_I(t)$, appear to be too small to magnify company level diminishing returns into substantial enough returns to scale at the macro level.[18]

Pro-cyclicality is a feature of most micro and macro data sets that have been reported in the literature, and a variety of explanations have been advanced to account for it.[19] Probably the most plausible alternative to economies of scale and/or external effects is measurement errors which arise from an inability accurately to observe capital utilisation, hours variation, labour hoarding, changing levels of effort or variations in the quality of labour and capital inputs properly. However, market power, capital market constraints and/or union power or industrial relations practices may vary over the cycle in a way which produces the same effects on measured productivity growth.[20] Our data are not rich enough to explore any of these alternatives, but we suspect that the mismeasurement of inputs (i.e. failing to correct cyclical variations in effort or capacity utilisation) may be a more important part of the story than competitive pressures or capital market constraints. The important question for our purposes, however, is whether these kinds of measurement errors create bias in estimating the degree of returns to scale. Equation (3) basically describes a regression of input growth and aggregate output growth on each firm's output growth with the following interpretation: the coefficient on the first term measures internal economies of scale, while the second measures external economies. However, the second term might also act as a proxy for the wide range of other factors which could make company productivity growth vary pro-cyclically, and their omission might, therefore, bias estimated values of ϕ upwards. However, this potential bias works towards reinforcing our basic conclusions, namely that agglomeration economies are a quantitatively unimportant component of company productivity growth.

4.2 Instrumental variables

Simultaneity bias is a problem which has caused some unease in studies of productivity growth, and has led some to prefer calculating the Solow residual directly (using observed factor shares) rather than through estimation.[21] Three different sources of bias might exist: first, since the aggregate variable in (3) is the sum of its micro parts, it may be correlated with the residual; second, if input and output choices occur simultaneously after the exogeneous shocks are realised, then the input growth term in (3) will be correlated with the residual; and, third, the lagged dependent variable may be correlated with the residual if the latter is not a white noise process. The first problem is not, we think, serious. The output of individual firms is a very small part of aggregate output, and we have used *net* industry and *net* aggregate output measures. The results also go through basically unaltered if we use gross industry or aggregate real output growth measures rather than the sum of the output growth rates of the firms in our sample (as in table 10.1).[22] The second problem also may not be too serious: the simultaneous choice of inputs and outputs ought

Table 10.2 *Alternative estimates of regression (i)*

Specification/β	Δx	$\Delta y_i(t-1)$	Δy_I	Δy aggregate	R^2	Diagnostics
(a) 1976–95 sales						
Fixed effects	0.6398 (0.024)	0.0614 (0.016)	0.2629 (0.046)	0.8047 (0.110)	0.2536	χ^2_4=80.37 (0.000)
Robust regression	0.4523 (0.014)	0.1334 (0.010)	0.2062 (0.026)	0.6346 (0.066)	N/A	$_{4,3537}$=427.87 (0.000)
Single constant	0.6601 (0.060)	0.1082 (0.031)	0.2288 (0.036)	0.8011 (0.106)	0.2556	$_{326,3211}$=0.590 (1.000)
2SLS (ldv)	0.5275 (0.050)	0.3561 (0.062)	0.2297 (0.034)	0.7429 (0.105)	0.2063	χ^2_5=12.478 (0.014)
2SLS (ldv & inputs)	0.5209 (0.246)	0.3236 (0.103)	0.2325 (0.041)	0.7416 (0.145)	0.2193	χ^2_6=28.177 (0.000)
2SLSDV (ldv)	0.5248 (0.056)	0.3157 (0.070)	0.2676 (0.044)	0.7398 (0.111)	0.2559	χ^2_5=12.918 (0.012)
2SLSDV (ldv & inputs)	0.8149 (0.227)	0.2020 (0.065)	0.2505 (0.049)	0.8387 (0.137)	0.2633	χ^2_6=26.719 (0.000)
AH (ldv)	0.2339 (0.041)	0.5568 (0.111)	0.0145 (0.014)	0.0473 (0.054)	N/A	χ^2_1=70.514 (0.000)
AH (ldv & inputs)	−0.2785 (0.076)	0.6929 (0.093)	0.0278 (0.016)	0.1145 (0.054)	N/A	χ^2_2=73.178 (0.000)
GMM (ldv)	0.1695 (0.004)	0.3126 (0.001)	0.0182 (0.002)	−0.3095 (0.008)	0.0896	m_2=−0.629 (0.529)
GMM (ldv & inputs)	−0.0247 (0.003)	0.3308 (0.001)	0.0238 (0.002)	−0.2584 (0.006)	0.0160	m_2=−0.466 (0.642)
(b) 1976–82 sales						
Fixed effects	1.0231 (0.056)	−0.1751 (0.025)	0.2037 (0.060)	0.1571 (0.127)	0.2279	χ^2_4=196.11 (0.000)
Robust regression	0.9058 (0.035)	−0.0284 (0.017)	0.2315 (0.033)	0.1322 (0.077)	N/A	$_{4,2101}$=239.58 (0.000)
Single constant	1.0125 (0.138)	−0.0312 (0.035)	0.1977 (0.053)	0.1047 (0.106)	0.2442	$_{490,1612}$=0.987 (0.567)
2SLS (ldv)	0.8524 (0.145)	0.2456 (0.108)	0.1556 (0.053)	0.2481 (0.109)	0.2186	χ^2_4=19.784 (0.001)
2SLS (ldv & inputs)	0.3129 (0.286)	0.3270 (0.134)	0.2425 (0.070)	0.2932 (0.119)	0.1341	χ^2_5=36.696 (0.000)
2SLSDV (ldv)	0.9565 (0.194)	0.1505 (0.126)	0.1890 (0.091)	0.3937 (0.205)	0.3931	χ^2_4=12.051 (0.017)
2SLSDV (ldv & inputs)	2.0283 (0.906)	0.0159 (0.091)	0.1256 (0.101)	0.2430 (0.261)	0.3123	χ^2_5=25.784 (0.000)
AH (ldv)	1.5541 (0.405)	−0.8811 (0.409)	0.1618 (0.103)	−1.2643 (0.448)	N/A	χ^2_1=0.377 (0.539)
AH (ldv & inputs)	−1.184 (1.229)	0.7277 (0.413)	0.6480 (0.293)	0.7040 (0.766)	N/A	χ^2_2=16.896 (0.000)
GMM (ldv)	0.7707 (0.072)	0.0215 (0.022)	0.2927 (0.046)	0.1984 (0.096)	0.2587	m_2=−1.568 (0.117)
GMM (ldv & inputs)	0.6515 (0.082)	0.0282 (0.021)	0.3419 (0.048)	0.0389 (0.090)	0.2433	m_2=−1.705 (0.088)

Table 10.2 (continued)

Specification/β	Δx	Δy$_i$(t–1)	Δy$_I$	Δy aggregate	R^2	Diagnostics
(c) 1983–95 sales						
Fixed effects	0.7749 (0.033)	0.0629 (0.023)	0.0549 (0.018)	0.3102 (0.059)	0.3260	χ^2_4=58.70 (0.000)
Robust regression	0.6056 (0.017)	0.1374 (0.012)	0.0463 (0.009)	0.2052 (0.031)	N/A	$_{4,1691}$=414.24 (0.000)
Single constant	0.7857 (0.088)	0.1235 (0.046)	0.0564 (0.020)	0.3282 (0.059)	0.3290	$_{184,1507}$=0.711 (0.998)
2SLS (ldv)	0.7145 (0.106)	0.3465 (0.018)	0.0284 (0.059)	0.1816 (0.080)	0.3028	χ^2_4=14.716 (0.005)
2SLS (ldv & inputs)	1.6542 (0.423)	0.0149 (0.105)	0.0249 (0.023)	0.1531 (0.068)	0.0503	χ^2_5=13.395 (0.020)
2SLSDV (ldv)	0.7596 (0.090)	0.3398 (0.141)	0.0243 (0.022)	0.1580 (0.065)	0.3691	χ^2_4=10.525 (0.032)
2SLSDV (ldv & inputs)	1.7315 (0.512)	0.1320 (0.101)	0.0136 (0.031)	0.1206 (0.077)	0.1342	χ^2_5=10.046 (0.074)
AH (ldv)	0.8702 (0.109)	0.3078 (0.108)	0.0367 (0.023)	0.1072 (0.067)	0.3106	χ^2_1=14.059 (0.000)
AH (ldv & inputs)	0.3641 (0.303)	0.4468 (0.103)	0.0358 (0.022)	0.1708 (0.087)	0.1743	χ^2_2=51.298 (0.000)
GMM (ldv)	0.7521 (0.015)	0.2007 (0.005)	0.0354 (0.007)	0.1424 (0.027)	0.3352	m_2=–0.588 (0.556)
GMM (ldv & inputs)	0.6061 (0.014)	0.2361 (0.003)	0.0128 (0.006)	0.1989 (0.023)	0.3128	m_2=–0.565 (0.572)
(d) 1983–91 value added						
Fixed effects	0.8836 (0.036)	0.0144 (0.028)	0.0342 (0.017)	–0.0081 (0.042)	0.4865	χ^2_4=9.48 (0.050)
Robust regression	0.6828 (0.018)	0.1224 (0.014)	0.0375 (0.009)	0.0410 (0.022)	N/A	$_{4,1028}$=517.24 (0.000)
Single constant	0.8833 (0.068)	0.0575 (0.056)	0.0303 (0.013)	–0.0002 (0.042)	0.4881	$F_{184,844}$=0.625 (1.000)
2SLS (ldv)	0.8607 (0.065)	0.0683 (0.058)	0.0339 (0.014)	0.0226 (0.036)	0.4818	χ^2_4=12.894 (0.012)
2SLS (ldv & inputs)	1.2258 (0.284)	–0.0511 (0.117)	0.0304 (0.016)	0.0234 (0.039)	0.4214	χ^2_5=6.845 (0.077)
2SLSDV (ldv)	0.8342 (0.095)	0.0140 (0.077)	0.0368 (0.018)	0.0338 (0.044)	0.5652	χ^2_4=16.114 (0.003)
2SLSDV (ldv & inputs)	1.3603 (0.358)	–0.0771 (0.095)	0.0239 (0.023)	0.0410 (0.050)	0.4639	χ^2_5=5.901 (0.117)
AH (ldv)	0.7817 (0.109)	–0.0628 (0.205)	0.0284 (0.015)	–0.1534 (0.046)	0.4095	χ^2_1=16.807 (0.000)
AH (ldv & inputs)	0.8484 (0.170)	0.0791 (0.162)	0.0279 (0.016)	–0.2095 (0.057)	0.4073	χ^2_2=36.545 (0.000)
GMM (ldv)	0.6628 (0.025)	0.1010 (0.019)	0.0312 (0.011)	–0.0994 (0.027)	0.4286	m_2=–2.138 (0.033)
GMM (ldv & inputs)	0.4566 (0.029)	0.1409 (0.020)	0.0404 (0.008)	–0.0589 (0.019)	0.3843	m_2=–1.745 (0.081)

Notes to Table 10.2

1. Heteroscedasticity consistent standard errors in parentheses below estimated coefficients for *single constant*, *2SLS/DV* and *AH* estimators.
2. *Robust regression* is an iteratively reweighted (Huber then biweight) least squares estimator on data screened for gross outliers using Cook's distance $D > 1$ with standard errors calculated using the pseudovalues approach (see Street, Carroll and Ruppert, 1988).
3. *2SLSDV* is a two stage, least squares, dummy variable estimator. Instruments used in all two stage, least squares estimates are once lagged first differences in the log of employment, real replacement cost of the capital stock, real cash flow, market capitalisation, total new fixed assets (investment), total gross plant and machinery and total gross land and buildings.
4. *AH* is an Anderson–Hsiao instrumental variable estimator. Instruments used for the lagged dependent variable are
$Z_i = [\mathrm{diag}(y_{i(t-2)}, \Delta y_{i(t-2)}, x_{i(t-1)}, \Delta x_{i(t-1)}) \vdots (\Delta \overline{x}_{iT})']$
(losing one cross-section) and for the inputs
$Z_i = [\mathrm{diag}(y_{i(t-2)}, \Delta y_{i(t-2)}, x_{i(t-1)}, \Delta x_{i(t-1)}) \vdots (\Delta \overline{x}_{iT})']$
where x is defined as in equation (x.2) and $\Delta \overline{x}$ are the remaining exogenous independent variables. These instruments are valid in the absence of serial correlation in the untransformed, levels model (i.e. MA(1) errors in the transformed, first differences model).
5. *GMM* is a two step generalised method of moments estimator. The instrument set for the lagged dependent variable is of the form
$Z_i = [\mathrm{diag}(y_{i1}, \ldots, y_{i(T-2)}) \vdots (\Delta x'_{i3}, \ldots, \Delta x'_{iT})']$;
that for the inputs in addition to the lagged dependent variable
$Z_i = [\mathrm{diag}(y_{i1}, \ldots, y_{i(T-2)}, x_{i1}, \ldots, x_{i(T-1)}) \vdots (\Delta \overline{x}_{i3}, \ldots, \Delta \overline{x}_{iT})']$
where for companies with less than the maximum number of observations the rows of Z_i corresponding to the missing equations are deleted and the missing values of y_i (x_i) in the remaining rows are replaced by zeros. These instruments are valid in the absence of serial correlation in the untransformed, levels model (i.e. MA(1) errors in the transformed, first differences model).
6. Characters in parentheses after *2SLS/DV*, *GMM* and *AH* estimators indicate which regressors are instrumented.
7. Diagnostics (p values in parentheses) for *fixed effects* regressions refer to Hausman tests for random effects (H0: unsystematic coefficient differences between random and fixed effects estimators); for *robust regressions* to F tests on all slope coefficients; for *single constant* regressions to F tests on fixed effects (H0: fixed effects can be simplified to a single constant); for *2SLS/DV* and *AH* regressions to Sargan tests of the overidentifying restrictions on the instruments (H0: excluded instruments uncorrelated with second stage errors); and for *GMM* regressions, a test for second order serial correlation (H0: serial independence) described in Arellano and Bond (1991).
8. For *GMM* regressions, tests for first order serial correlation described in Arellano and Bond (1991) are not presented but the null hypothesis (serial independence) cannot be rejected in the specifications on panels (b) and (d) only.

to bias up our estimates of internal scale economies, but our estimates of these economies are already fairly low (and possibly implausibly so). The third problem is more of a cause for concern.

In order to address potential biases we need to find some instruments that might alleviate the problem. Table 10.2 displays a number of different experiments. The GMM estimates when the lagged dependent variable is instrumented are valid, as are the GMM estimates where both the lagged dependent variable and the inputs are instrumented. Note that the R^2 in all of the GMM(ldv) regressions is greater than in all of the GMM(ldv&inputs) regressions. As serial correlation tests show that the instruments are valid in both cases, the former seem to be more efficient than the latter, and we are inclined to conclude that instrumenting inputs is not necessary. The AH estimates and 2SLS estimates always fail the Sargan instrument validity test (although the AH tests which instrument just the lagged dependent variable do not always fail). We suspect that first-order serial correlation is induced by differencing, and hence only instruments dated before t–2 will be valid. If there is, in addition, serial correlation in the levels model, then only instruments dated earlier than t–3 will be valid. However, the 2SLS estimates and the GMM and AH estimates which instrument inputs contain t–1 terms, and these are probably the source of the problem. It follows that the AH and GMM estimates on the longest panels are likely to be the most reliable.

The estimates of returns to scale produced by the GMM(ldv) and AH(ldv) regressions vary across samples pretty much in the same way that all of the estimates vary sample by sample. Generally, they are consistent with diminishing returns, although panel (a) displays some implausibly low estimates, while those on panel (b) seem rather high. Much more interestingly, estimates of industry and macro effects are always very small in these regressions, suggesting that if there is a serious simultaneity bias on table 10.1, it is that the estimates of ϕ_1 and ϕ_2 are biased upwards; i.e. that agglomeration economies are exaggerated in the baseline estimates on table 10.1. Interestingly, our estimates of spillovers are also a little sensitive to the use of instruments, and, in some of our instrumental variables estimates, spillovers became (marginally) significant. We have no good explanation for this.

4.3 Sample composition

A lot of evidence has accumulated over the years suggesting that data series recording productivity growth rates are liable to structural breaks and that heterogeneities between firms may be present. Our baseline sample extends from 1974 to 1990 and involves 216 firms. We have already reported on experiments that use only the first fourteen years of the data (see table 10.1), and table 10.2 shows further calculations which also involve changing sample sizes. Using panel (a) as a benchmark vis-à-vis panels (b) and (c), it seems plain that returns to scale and the two coefficients associated with agglomeration effects are possibly slightly higher in the period 1976–82 than in the period 1983–95, but the real curiosity is that the estimates for the 1976–95 pooled sample are lower than for either of the sub-samples.[23] It seems possible that some structural change (possibly associated with the recession in 1981 and 1991) occurred during the 1976–95 period, but quite why it should have this effect on returns to scale is unclear. We suspect that the culprit may be variations in sample size, meaning that the pooled sample may (for some unknown reason) contain fewer firms which enjoy substantial scale economies. This degree of temporal instability or heterogeneity makes us a little uneasy with the kind of large-scale panel data estimates of productivity growth equation which we have reported here, even though our qualitative conclusions seem to be reasonably robust to the instability.

This leads to a second, related issue. The way that agglomeration effects and spillovers have been modelled in (7) and (9) may be misleading: external effects may have a slower, more secular effect on productivity that is hard to fix with accuracy in a short time period, and agglomeration effects (or technology spillovers) may not be driven by high frequency changes in $\Delta y_I(t)$ and $\Delta y(t)$. One way to address this concern is to estimate (7) and (9) using long first differences. We have done this using a balanced panel drawn from the sample used in table 10.1. The balanced panel estimates of regressions (i)–(iii) are very similar to those shown in table 10.1. The long first differences yield slightly higher estimates of returns to scale, but both agglomeration coefficients were much smaller and quite insignificant. Roughly the same applies to the samples used in table 10.2: long differences on balanced panels yielded very small, insignificant agglomeration effects.

A third concern with our results is the fact that sample selection may lead to bias. In particular, if firms which enjoy large internal or external economies of scale (or are more innovative) are more likely to survive, then our estimates of scale economies will probably be biased upwards. In fact, the main cause of the disappearance of firms from our sample (most of whom are quoted) is mergers or takeovers, and firm size (and short-term financial health) is the dominant selection criterion. Still, we replicated a number of these regressions on balanced panels, and observed no major differences in the estimates. This is a fairly common observation in panels of this kind, and we are inclined to believe that sample attrition may not be too serious a problem.

Fourth and finally, the distribution of productivity growth rates across firms has both a high variance and a very wide range. This raises the possibility that our estimates might be sensitive to outliers. The estimates of table 10.2 include robust regression estimates which try to correct for this. In each case, the robust estimate of scale economies is slightly smaller than either the fixed effects of single constant estimate, and the coefficient on $\Delta y(t)$ is larger (in three of the four panels). No obvious effect was observed on the estimates of the effects of the innovations, patents or technology spillovers on productivity growth. We surmise from all of this that the outliers in our data are firms whose output growth rates are much more variable than input growth rates, possibly because they are unusually sensitive to variations in macroeconomic growth rates. That is, the outliers in our data may be the firms most affected by agglomeration economies.

4.4 Measurement errors

Measurement errors are probably the single most important source of concern in productivity studies, and they can have a number of (sometimes) quite unpredictable effects on the numbers. We have managed to explore three particular potential sources of such problems in our data:

First, we examined the sensitivity of our results to the details of how capital stock was measured. All of the regression shown on table 10.2 use a different measure of capital stock from that shown on table 10.1. The major difference between the two calculations of capital stock is, we believe, that the estimates used in the baseline sample will be more pro-cyclical (because of the way that scrapping is handled).[24] However, we recalculated capital stock using the second method, and the two series are highly correlated. Further, the results displayed on table 10.2 suggest that our results are not very sensitive to this difference.

Second, and of more worry, omitting materials is likely to exacerbate the apparent pro-cyclicality of the Solow residual, since materials are generally a variable input whose usage is unlikely to be affected by variations in effort.[25] At the very least, its omission is likely to bias upwardly estimated coefficients on the other variable inputs (like labour), exaggerating the degree of scale economies. For many, the right answer to this potential problem is to estimate a value added production function (as we did in table 10.1). The two problems which this 'solution' creates are: the assumption on factor substitution which turns an ordinary production function into a value added production function are rather strong, and good data on value added at company level are not available in the UK.[26] The first three panels of table 10.2 show regressions in which real output rather than value added is the dependent variable. The interesting comparison is between panels (c) and (d), and the main result induced by the change in the dependent variable is that estimates of returns to scale seem to be just slightly lower and agglomeration effects are slightly higher when real output data are used than when value added data are used. No effects on technology spillovers were observed when we recomputed the regressions on table 10.1 with real output rather than value added.

Third, and finally, fixed effects sometimes capture important omitted variables which have a relatively permanent effect on the dependent variable. Table 10.2 shows that a fixed effects specification is always preferred to a random effects specification, but that the fixed effects can usually be simplified to a single constant. This is not a surprise given the relative unimportance of between variation in the data. What is more, there seems to be little practical difference between fixed effects and single constant equation results. The inclusion of fixed effects also had no discernible effect on the estimates of technology spillovers.

4.5 Preliminary conclusions

There seem to be two conclusions to draw from all of this. First, most of the traditional sources of concern suggest that our estimates of both internal and external economies of scale are likely to be biased upwards. Since our view is that neither type of scale economy is a major driver of movements in company productivity growth, these problems do not really undermine our confidence in the conclusions. Second, in practice the broad qualitative conclusions that we drew at the end of section 2 appear to be fairly robust, although there is clearly a lot of variation in some of the point estimates in the different regressions. Of most concern, in our view, is the temporal instability in the estimates (which is a feature of many other data sets), since it seems clear that changes in the parameters of equations (5) and (9) are being driven by omitted determinants of productivity growth that may be easy to confound with returns to scale or agglomeration effects.

5 Conclusions

It is possible to tell many interesting and rather plausible stories about the importance of external effects on the performance of particular firms.

There is no doubt that the success of particular firms sometimes rests upon an infrastructure of publicly and privately provided knowledge and skills. Further, many firms, when taken together with some of their rivals, buyers and suppliers, realise a performance that is greater than the sum of these parts. Our data show that there is a noticeable and very systematic discrepancy between returns to scale measured at company, industry and aggregate level which might be consistent with the presence of external effects. This said, it seems to be the case that variations in total factor productivity across companies and over time do not seem to be driven in any substantively important way by external effects, although the evidence does suggest that some companies in some sectors do derive stronger benefits than others. Probably more surprisingly, innovation, patenting and technological spillovers also do not seem to be very precisely correlated with cross-section or times-series variations in company productivity performance.

The data suggest that company productivity growth and the production of major innovations or (somewhat less clearly) patents are highly idiosyncratic, and what happens to different companies year in and year out is only very weakly related to what happens to other companies. Most of the within- (i.e. year-to-year) variation in productivity growth is random, meaning that productivity levels (roughly speaking) follow a random walk. Much the same is true of innovative output like patents and the production of 'major' innovations, and few firms manage to generate major innovations or patents in spells that last longer than a single innovation or a couple of patents produced in a single year. Company performance, particularly when measured by productivity growth rates or innovative output, is pretty erratic over time.

Appendix I: The baseline sample

The firms are all drawn from the EXSTAT data file (the industry data was matched with the firm data using the EXSTAT industry codes). This covers the period 1972–90 and consists of 4082 firms. Selecting only those firms which are in manufacturing reduces this to 1044. Of these 1044 firms, 545 had the necessary data to calculate all the firm variables, but only 216 of these have eight or more consecutive observations after taking first differences and allowing for a lag.

The principal variables are as follows: *Value-Added*: the sum of the cost of employees, profits before tax, the depreciation charge and interest payments. It is deflated by industry specific price indices. (Mean = 34,632, standard deviation (between and within components) = 70,600 (69,625, 24,821).) *Employment and Wage Bill*: number of employees and cost of employees. (Mean employment = 2790, standard deviation (between and within components) = 5122 (5076, 1583)); (mean wage bill = 22,754, standard deviation (between and within components) = 51,870 (48,439, 23,662).) *Patents and Innovations*: Annual number of patents granted to each firm by the US Patents Office and major innovations produced – kindly provided by SPRU at the University of Sussex. (Mean patents = 0.221, standard deviation (between and within components) = 1.186 (0.993, 0.749);) (mean innovations = 0.031, standard deviation (between and within components) = 0.253 (0.127, 0.205).) *Two-Digit Industry Output and Manufacturing Output*: Gross value-added (from the *Census of Production* summary tables), deflated by industry specific price indices and by the manufacturing price index respectively. The producer price indices used for this were taken from the *Annual Abstract of Statistics*. (Mean industry output = 5,111,983, standard deviation = 2,281,267; mean manufacturing output = 739,218,100, standard deviation = 79,079,270.) *Capital Stock*: The firm capital stock figures were calculated using the method proposed by Wadhwani and Wall (1986). This is based on the following balance sheet identity on the movement in a firm's nominal assets:

$$HCK_t = HCK_{t-1} + AD_t + DISP_t + NSC_t + SDISP_t + REV_t + CC_t + OTH_t$$

where HCK_t is the historical cost of the firm's assets at the end of period t, AD_t is additions, $DISP_t$ is disposals, NSC_t is the historical cost of assets of new subsidiaries purchased, $SDISP_t$ is the historical cost of assets of subsidiaries disposed of, REV_t is revaluations, CC_t is currency changes and OTH_t is 'other' during period t (assets in the 'other' category have either just been purchased or are just about to be sold so they are treated as either additions or disposals depending upon whether the figure for 'other' is positive or negative). The current real gross capital stock is estimated by calculating the real value of the terms on the right-hand side of this identity (as companies do not treat revaluations and currency charges in a consistent manner these two terms are excluded from the analysis). Both additions and new subsidiaries are valued in terms of current prices. Therefore we estimated the real value of them by multiplying these terms by the ratio of the current price of investment goods to the price in the base year – the year in terms of which the real capital stock is being measured. (Mean capital stock = 95,081, standard deviation (between and within components) = 208,471 (193,766 , 100,204).)

Calculating the real value of disposals and the sale of subsidiaries was more complicated as these assets are recorded at historical cost (that is, at the price ruling when the assets were purchased). Therefore an assumption had to be made about the age of these assets. We experimented with the same assumptions that Wadhwani and Wall used, namely eight years, the average age of the firm's assets and the service life of the firm's assets. The average age and service life of a firm's assets were approximated by the ratio of current depreciation to accumulated depreciation and to the current historical cost of the firm's assets respectively. We found that while using the average age and eight years produced very similar estimates of the growth rate of the capital stock, using the service life produced estimates which implied that there had been large-scale and widespread capital scrapping throughout the whole of the 1980s. We felt this amount of capital scrapping was implausible and, therefore, used the average age estimates in this chapter. Finally, to take account of assets which were financially leased in the period prior to 1981, the real value of the figure for plant hire given in the firm accounts was capitalised using a real interest rate of 5 per cent (since 1981 companies have had to treat financially leased assets as if the company actually owed them).

Thus, we estimated current real gross capital stock of each firm using the following equation:

$$RV_t = RV_{t-1} + (AD_t + NSC_t) * \left(\frac{P_0}{P_t}\right) - (DISP_t + SDISP_t * \left(\frac{P_0}{PA_t}\right) + FA_t$$

where RV_t is the real gross capital stock valued at base year prices P_0, P_t is the current price, A_t is the current average age of the firm's assets and FA_t is the current real value of financially leased assets in the period prior to 1981. An initial figure for real gross capital stock was estimated using the following equation:

$$RV_1 = HCA_1 * \left(\frac{P_0}{PA_t}\right)$$

Appendix II: The other samples

A change in UK legal requirements in July 1982 concerning the reporting of employment and wages in company accounts means only a small number of companies continued to report these data after that date on a basis which was consistent with that used previously. For consolidated reports and accounts of companies with overseas subsidiaries prior to July 1982, only disclosure of domestic employment and wages was mandatory. After that date, however, only disclosure of total employment and wages was mandatory. Few companies voluntarily continued to disclose domestic employment and wages after 1982. Hence, the samples used for alternative estimates of regression (i), given in table 10.2, are roughly speaking split at that date. Reported sales are total, not domestic, throughout the period.

The 1976–82 sample with sales as the dependent variable (the 1976–82 'sales sample') used in panel (b) of table 10.2 is essentially the unbalanced panel used in Geroski, Machin and Walters (1997) and is described in more detail there. It forms the intersection of three large, secondary panel datasets: the DataStream online UK firms' accounts database, the Science Policy Research Unit (SPRU, University of Sussex) Innovations database and the SPRU US Patents database. Of the 615 firms, 1976–82, used in that paper, 491 had the necessary employment data to calculate $\Delta x_i(t)$ in regression (i) and had at least five observations of our principal variables. Fifty of these firms have five observations, 252 have six observations and 189 have seven observations. Of these 491 firms, 327 continued to report domestic employment in a consistent manner until 1995, and these 327 are the 1976–95 sales sample used in panel (a) of table 10.2. The distribution of T_i for this unbalanced panel is min = 6, 5 percentile = 7, 25 percentile = 8, median = 11, 75 percentile = 19, 95 percentile = 20, max = 20. The panel is as follows: 7 firms have 6 observations, 47 firms have 7 observations, 31 firms have 8 observations, 38 firms have 9 observations, 18 firms have 10 observations, 25 firms have 11 observations, 8 firms have 12 observations, 17 firms have 13 observations, 12 firms have 14 observations, 9 firms have 15 observations, 10 firms have 16 observations, six firms have 17 observations, 13 firms have 18 observations, 55 firms have 19 observations and 31 firms have 20 observations.

We were concerned that the switch from using real valued added to real sales as the dependent variable in moving from the estimate of regression (i) on table 10.1 to the estimates of regression (i) on panels (a) and (b) of table 10.2 would invalidate our robustness conclusions, but very few of our 491 (327) firms have sufficient data prior to 1982 to calculate value added. Moreover, and as previously mentioned, few firms report domestic employment and wage data after 1982. This meant we were unable to calculate value added for these firms using domestic wage data after 1982 and unable to calculate value added using other components before 1982. Therefore, we considered 186 of these 327 firms who (1) reported employment in a manner consistent with the July 1982 change in UK legal requirements from 1983–95 (that is, total employment and wages) and (2) had coverage of other data sufficient to calculate value added from 1983–91. These are the 186 firms in the 1983–91 value added sample used in panel (d) of table 10.2 and the 1983–95 sales sample used in panel (c) of table 10.2 (estimates reported in panel (c) of table 10.2 are thus included for comparability with the dependent variable used in the samples in panels (a) and (b)). 15 of these firms have 5 observations, 23 have 6 observations, 48 have 7 observations, 37 have 8 observations and 63 have 9 observations.

Our principal variables are as follows:

Sales (DataStream item 104) deflated by 2-digit 1980 SIC industry specific deflators (taken from the *Annual Abstract of Statistics*). This is the amount of goods and services to third parties, relating to the normal activities of the company and does not include Value Added Tax (or other turnover taxes) or trade discounts. The mean of this variable for the 1976–82 sample $m_{76-82} = 170{,}452$, its standard deviation (between and within components) $s_{76-82} = 485{,}738$ ($s_{b,76-82} = 473{,}172$, $s_{w,76-82} = 63{,}140$) and for the other samples $m_{76-95} = 206{,}728$, $s_{76-95} = 637{,}076$ ($s_{b,76-95} = 569{,}127$, $s_{w,76-95} = 127{,}220$) and $m_{83-95} = 223{,}449$, $s_{83-95} = 727{,}586$ ($s_{b,83-95} = 704{,}913$, $s_{w,83-95} = 112{,}567$).

Value Added computed as cost of employees + pre-tax profits + depreciation + interest payments (DataStream items 117 + 400 + 338 + 153) deflated by 2-digit 1980 SIC industry specific deflators (taken from the *Annual Abstract of Statistics*). The mean of this variable for the 1983–91 sample $m_{83-91} = 127{,}383$, its standard deviation (between and within components) $s_{83-91} = 480{,}452$ ($s_{b,83-91} = 457{,}486$, $s_{w,83-91} = 49{,}663$).

Domestic Employment (DataStream item 216) including part time. Figures available vary between number employed at the year end and average number employed throughout the year. The mean of this variable for the 1976–82 sample $m_{76-82} = 5295$, its standard deviation (between and within components) $s_{76-82} = 13{,}608$ ($s_{b,76-82} = 13{,}526$, $s_{w,76-82} = 1726$) and for the other sample $m_{76-95} = 4906$, $s_{76-95} = 13{,}029$ ($s_{b,76-95} = 13{,}049$, $s_{w,76-95} = 3{,}658$).

Total Employment (DataStream item 219) including part time. Figures available vary between number employed at the year end and average number employed throughout the year. The mean of this variable for the 1983–91 sample $m_{83-91} = 6820$, its standard deviation (between and within components) $s_{83-91} = 17,843$ ($s_{b,83-91} = 17,975$, $s_{w,83-91} = 2259$) and for the other sample $m_{83-95} = 6,486$, $s_{83-95} = 16,546$ ($s_{b,83-95} = 17,385$, $s_{w,83-95} = 2996$). Capital Stock deflated using an aggregate fixed manufacturing investment implied deflator (taken from annual editions of the Blue Book).

Capital stock is calculated using the perpetual inventory method (see Blundell, Bond and Schiantarelli, 1992) since for most of the years of the data, accounting rules did not require replacement cost to be declared. The mean of this variable for the 1976–82 sample $m_{76-82} = 113,663$, its standard deviation (between and within components) $s_{76-82} = 450,899$ ($s_{b,76-82} = 434,145$, $s_{w,76-82} = 36,171$) and for the other samples: $m_{76-95} = 152,609$, $s_{76-95} = 563,950$ ($s_{b,76-95} = 482,299$, $s_{w,76-95} = 96,149$); $m_{83-95} = 171,672$, $s_{83-95} = 612,537$ ($s_{b,83-95} = 555,844$, $s_{w,83-95} = 82,851$) and $m_{83-91} = 164,665$, $s_{83-91} = 601,762$ ($s_{b,83-91} = 561,771$, $s_{w,83-91} = 75,789$).

Notes

1 We are obliged to the Gatsby Foundation for support, and to Jonathon Haskel, Bronwyn Hall and Zvi Griliches for helpful comments on an earlier draft. The usual disclaimer applies. Ian Small moved to the Bank of England during the course of this project, and, needless to say, the Bank is not in any way responsible for, or necessarily in agreement with, the contents of the chapter.
2 This selection method generated 2480 firm-year observations to work with (68 per cent of the 3672 observations potentially available from a panel of this size). The structure of the panel is as follows: 34 firms had 8 years of data, 27 had 9 years, 35 had 10 years, 29 had 11 years, 31 had 12 years, 13 had 13 years, 8 had 14 years, 6 had 15 years, 8 had 16 years, and 25 firms were observed continuously over the whole 17-year period.
3 Hall (1990) shows that the two explanations of why $\gamma > 1$ (i.e. increasing returns and market power) can be distinguished if a is measured as labour's share of total costs. In this case, $\gamma = 1$ if returns to scale are constant but firms have market power, while $\gamma > 1$ if increasing returns are present but firms have no market power. For empirical work on returns to scale and monopoly power based on this observation, see Hall, (1988), Roeger (1995), Chirinko

and Fazzari (1994), Bottasso and Sembenelli (1997) and others.
4 See Davis, Haltiwanger and Schuh (1996), Geroski and Gregg (1997) and others.
5 See Murphy, Schliefer and Vishney (1989), Schliefer (1986), Bresnahan and Trajtenberg (1995), Diamond (1982), Hall (1991) and others.
6 See Porter (1990), Krugman (1991), Saxenian (1994), Arthur (1994) and others. For some recent econometric evidence on the effects of employment density (which may create local increasing returns) on productivity differences between US states, see Ciccone and Hall (1996).
7 This argument was made by Griliches (1979) in the context of technology spillovers; the text follows Caballero and Lyons (1990, 1992).
8 See also Nickell, Wadhwani and Wall (1992), who argue that *'long run decreasing returns are inherently most unlikely'* (p. 1070), and are due to measurement error of inputs. This may also reflect the existence of U-shape cost curves (most of the firms in the sample are large) or a mix of increasing returns with decreasing economies of scope (most of them are multi-product). Notice that if firms have market power, then γ almost certainly overstates the degree of returns to scale (see note 3 above). Further, 'long-run' returns to scale are given by $\gamma/(1-\gamma)$ in (7), and the estimate of this is 0.68 in regression (i) on table 10.1, not 0.80.
9 Using (i), our point estimate of external economies is only just a little larger than those reported by Oulton (1996), who used data for 124 three digit UK manufacturing industries, 1954–86; see also Oulton and O'Mahony (1994). The original work using these techniques was reported by Caballero and Lyons (1990, 1992), who found that a 1 per cent rise in aggregate UK output would increase two-digit sector level input by about 0.26 per cent on average; see also Bartelsman, Caballero and Lyons (1994), who provide some evidence suggesting that linkages with customers may be an important part of the transmission of external effects between firms.
10 By contrast, using two-digit industry data for the US, Caballero and Lyons (1992), report an average value of $\Psi \approx 0.9$, which suggests that aggregation drastically reduces the variability in the underlying micro data.
11 See Griliches (1979, 1991) for a good discussion of the issues associated with measuring spillovers. Amongst other things, he points out that information flows between buyers and suppliers (i.e. vertical relationships) are not always (and perhaps not often) externalities. Most of the literature has tried to identify technology spillovers between horizontally related firms; Jaffee (1986), is one of the few studies which tries to identify a technological profile for each firm.
12 The imprecision of these estimates might be due to the infrequency in the occurrence of innovations (only 9 firms in this sample innovated over the period, producing an average of 0.039 innovations per firm-year). This is less true for patents (69 firms patented at least once over the sample period, producing 0.248 patents per firm-year). Patents are very heterogeneous in value (this is arguably somewhat less true for the data on major innovations) and this

undoubtedly increases the imprecision further.

13 Cohen and Levinthal (1989, 1990), describe this as a firm's 'absorptive capacity'; see also Malerba (1992). Evidence suggesting that innovative firms may absorb spillovers better than non-innovators is provided by Geroski, van Reenan and Walters (1993), who use a technique similar to that described in the text. They also observed that the profitability of innovating firms was relatively insensitive to macroeconomic fluctuations, they were noticeably more sensitive to spillovers than non-innovators.

14 For surveys of the empirical literature on spillovers, see Griliches (1991) and Nadiri (1993); Geroski (1985) is a more sceptical survey. Evidence that users may take a large share of the benefits of innovations relative to producers is presented in Scherer (1982), Griliches and Lichtenberg (1984), Geroski (1991) and others. For work on the effect of R&D on productivity growth, see Griliches and Mairesse (1983 and 1984), Harhoff (1994), Hall and Mairesse (1995), Wakelin (1997) and others.

15 See Geroski, Van Reenan and Walters (1997), who analyse the patent and innovation producing activities of firms in a sample similar to the baseline sample used here. They observed that the probability that a firm will innovate or patent in period $t+1$ given that it has done so in period t is very low, unless it has produced perhaps as many as 5 patents or 2–3 innovations in period t. Further, there is a rough pro-cyclicality to the incidence of innovation, and this may induce a rather mild positive correlation between the innovative activities of different firms which has little to do with technology spillovers; see Geroski and Walters (1995), Geroski, Machin and Walters (1997) for macro and micro based evidence.

16 The last four years of sample (3) were dropped because coverage of profits data becomes very patchy after that date.

17 We also re-ran the regressions on table 10.1 using the rates of growth of labour and capital entered separately, rather than combining them in the single factor share weighted term $\Delta y_i(t)$. This change has no substantial effects on estimates of returns to scale, agglomeration effects or technology spillovers.

18 It is also worth noting that the Ψ_i discussed in section 3 are based on values of $\theta_i(t)$ computed using observed factor shares (i.e. assuming constant returns to scale). If this is incorrect they should be positively (negatively) correlated with either the rate of growth of labour or of real capital if internal economies (diseconomies) of scale exist. In our data, these correlations are very small, which seems hard to reconcile with increasing returns to scale.

19 See the stimulating discussions in Hall (1990), Morrison and Berndt (1981), Burnside, Eichenbaum and Rebelo (1993), Fay and Medoff (1985), Bernanke and Parkinson (1991), Basu (1996) and others.

20 These variables are featured in recent work on UK company productivity growth. Nickell, Wadhwani and Wall (1992) and Nickell (1996) report work on company productivity growth in the UK which uses measures of debt and competitive pressures to help track variations in effort across firms and over time, and a recession shock variable plus a measure of overtime hours sug-

gested by Muellbauer (1984) to correct for cyclical effects. Their data strongly suggest decreasing returns at the company level, implying that other factors must be responsible for pro-cyclical productivity growth. Nickell, Nicolitsas and Dryden (1997) extends this work and also examines the role of shareholders. Coupled with Gregg, Machin and Metcalf (1993), these studies also report time varying affects associated with unionisation (see also Machin, 1991, for work on a sample of engineering firms). Andrews and Hughes (1993), Caves (1992), Mayes (1996), Dickerson, Geroski and Knight (1997) and others report similar work applied to industry data.

21 For a stimulating discussion of the issues, see Griliches and Mairesse (1995).

22 Caballero and Lyons (1990, 1992), and Oulton (1996), are (rightly) more concerned about the effects that correlations between $\Delta y(t)$ and the residual may have in their industry level work. They use aggregate inputs as an instrument for aggregate output. Oulton's results suggest that the aggregate inputs instrument leads to slightly higher estimates of agglomeration effects.

23 Recall that our innovations and patents data end in 1983 and 1988 respectively, so these comparisons can say nothing about whether the importance of technology spillovers varied over the sample period.

24 The capital stock figures used in table 10.1 follow Wadhwani and Wall's (1986) approach and estimate scrapping by revaluating the historical cost figures for scrapping in company accounts, while the figures used in table 10.2 follow Blundell, Bond and Schiantarelli (1992) and assume that a fixed proportion of the existing capital stock is scrapped each year irrespective of economic conditions. Assuming that firms increase their rate of capital scrapping in recessions/downturns and reduce it during upturns, then Wadhwani and Wall's approach will tend to produce more counter-cyclical estimates of capital scrapping and hence more pro-cyclical estimates of the capital stock, than the approach used by Blundell *et al.* Therefore, using this second set of capital stock figures in table 10.2 may produce larger estimates of the returns to scale coefficient (and also possibly in the coefficients on net two-digit output and net manufacturing output).

25 See Basu (1996), who uses this observation to devise a measure of the cyclical variation in input usage; see also Cuneo and Mairesse (1984), Hall and Mairesse (1994) and others for observations on the use of sales and real value added data in this context.

26 See Nickell, Wadhwani and Wall (1992). To get value added, we add pre-tax profits plus interest payments to the wage bill, but the profits data are net of bad debt, various other provisions, capital gains and, sometimes, are contaminated by current expensing of durable inputs.

References

Andrews, M. and Hughes, K. (1993), 'UK productivity in the 1980s: evidence

from a manufacturing four-digit panel', mimeo, University of Manchester.
Arellano, M. and Bond, S. (1991), 'Some tests of specification for panel data: Monte Carlo evidence and an application to employment equations', *Review of Economic Studies*, 58, pp. 277–97.
Arthur, B. (1994), *Increasing Returns and Path Dependence in the Economy*, Ann Arbor, University of Michigan Press.
Bartelsman, E., Caballero, R. and Lyons, R. (1994), 'Customer and supplier driven externalities', *American Economic Review*, 84, pp. 1075–84.
Basu, S. (1996), 'Pro-cyclical productivity: increasing returns or cyclical utilisation', *Quarterly Journal of Economics*, 104, pp. 719–51.
Bernanke, B. and Parkinson, M. (1991), 'Pro-cyclical labour productivity and competing theories of the business cycle: some evidence from interwar US manufacturing industries', *Journal of Political Economy*, 99, 3, pp. 439–59.
Blundell, R., Bond, S. and Schiantarelli, F. (1992), 'Investment and Tobin's Q: evidence from company panel data', *Journal of Econometrics*, 51, pp. 233–57.
Bottasso, A. and Sembenelli, A. (1997), 'Market power productivity and the EU single market programme', mimeo, Turin, CERIS-CNR.
Bresnahan, T. and Trajtenberg, M. (1995), 'General purpose technologies: engines of growth?', *Journal of Econometrics*, 65, pp. 83–108.
Burnside, C., Eichenbaum, M. and Rebelo, S. (1993), 'Labour hoarding and the business cycle', *Journal of Political Economy*, 101, pp. 245–73.
Caballero, R. and Lyons, R. (1990), 'Internal versus external economies in European industry', *European Economic Review*, 34, pp. 805–30.
 (1992), 'External effects in US pro-cyclical productivity', *Journal of Monetary Economics*, 29, pp. 209–25.
Caves, R. (1992) (ed.), *Industrial Efficiency in Six Nations*, Cambridge Mass., MIT Press.
Chirinko, R. and Fazzari, S. (1994), 'Economic fluctuations, market power and returns to scale: evidence from firm-level data', *Journal of Applied Econometrics*, 9, pp. 47–69.
Ciccone, A. and Hall, R. (1996), 'Productivity and the density of economic activity', *American Economic Review*, 86,1, pp. 54–70.
Cohen, W. and Levinthal, D. (1989), 'Innovation and learning: the two faces of R&D', *Economic Journal*, 99, pp. 569–96.
 (1990), 'Absorptive capacity: a new perspective on learning and innovation', *Administrative Science Quarterly*, 35, pp. 128–52.
Cuneo, P. and Mairesse, J. (1984), 'Productivity and R&D at the firm level in French manufacturing', in Griliches, Z. (ed.), *R&D, Patents and Productivity*, Chicago, University of Chicago Press.
Davis, S., Haltiwanger, J. and Schuh, S. (1996), *Job Creation and Destruction*, Cambridge, Mass., MIT Press.
Diamond, P. (1982), 'Aggregate demand management in search equilibrium', *Journal of Political Economy*, 90, pp. 881–94.
Dickerson, A., Geroski, P. and Knight, K. (1997), 'Productivity, efficiency and

strike activity', *International Review of Applied Economics*, 11, pp. 119–34.
Fay, J. and Medoff, J. (1985), 'Labour and output over the business cycle: some direct evidence', *American Economic Review*, 75, pp. 638–55.
Geroski, P. (1985), 'Do spillovers undermine incentives to innovate?', in Dowrick, S. (ed.), *Economic Approaches to Innovation*, Aldershot, Edward Elgar.
 (1991), 'Innovation and the sectoral sources of UK productivity growth', *Economic Journal*, 101, pp. 1438–51.
Geroski, P. and Gregg, P. (1997), *Coping with Recession*, Cambridge, Cambridge University Press.
Geroski, P., Machin, S. and Walters, C. (1997), 'Corporate growth and profitability', *Journal of Industrial Economics*, 45, pp. 171–89.
Geroski, P., Van Reenen, J. and Walters, C. (1993), 'The profitability of innovating firms', *Rand Journal of Economics*, 24, pp. 198–211.
 (1997), 'How persistently do firms innovate?', *Research Policy*, 26, pp. 33–48.
Geroski, P. and Walters, C. (1995), 'Innovative activity over the business cycle', *Economic Journal*, 105:431, pp. 916–28.
Gregg, P. Machin, S. and Metcalf, D. (1993), 'Signals and cycles: productivity growth and changes in union status in British companies 1984–1989', *Economic Journal*, 103, pp. 894–907.
Griliches, Z. (1979), 'Issues in assessing the contribution of R&D to productivity growth', *Bell Journal of Economics*, 10, pp. 92–115.
 (1991), 'The search for R&D spillovers', mimeo, Cambridge Mass., NBER.
Griliches, Z. and Lichtenberg, F. (1984), 'Inter-industry technology flows and productivity growth: a comment', *Review of Economics and Statistics*, 66, pp. 324–9.
Griliches, Z. and Mairesse, J. (1983), 'Comparing productivity growth: an exploration of French and US industrial and firm data', *European Economic Review*, 21, pp. 89–119.
 (1984), 'Productivity and R&D at the firm level', in Griliches, Z. (ed.), *R&D, Patents, and Productivity*, Chicago, University of Chicago Press.
 (1995), 'Production functions: the search for identification', mimeo, Harvard University.
Hall, B. and Mairesse, J. (1994), 'Comparing productivity of R&D in French and US manufacturing firms', mimeo, NBER.
 (1995), 'Exploring the relationship between R&D and productivity in French manufacturing firms', *Journal of Econometrics*, 65, pp. 263–93.
Hall, R. (1988), 'The relation between price and marginal cost in US industry', *Journal of Political Economy*, 96, pp. 921–47.
 (1990), 'Invariance properties of Solow's productivity residual', in Diamond, P. (ed.), *Growth/Productivity/Unemployment*, Cambridge Mass., MIT Press.
 (1991), 'Labour demand, labour supply and employment volatility', in *NBER Macroeconomics Annual*, 6, pp. 17–47.
Harhoff, D. (1994), 'R&D and productivity in German manufacturing firms', mimeo, University of Mannheim.
Jaffee, A. (1986), 'Technological opportunity and spillovers of R&D: evidence

from firms patents, profits and market values', *American Economic Review*, 76, pp. 984–1001.
Krugman, P. (1991), *Geography and Trade*, Cambridge Mass., MIT Press.
Machin, S. (1991), 'The productivity effects of unionisation and firm size in British engineering firms', *Economica*, 58, pp. 479–90.
Malerba, F. (1992), 'Learning by firms and incremental technical change', *Economic Journal*, 102, pp. 845–59.
Mayes, D. (1996) (ed.), *Sources of Productivity Growth*, Cambridge, Cambridge University Press.
Morrison, C. and Berndt, E. (1981), 'Short-run labour productivity in a dynamic model', *Journal of Econometrics*, 16, pp. 339–65.
Muellbauer, J. (1984), 'Aggregate production functions and productivity measurements', mimeo, Centre for Economic Policy Research.
Murphy, K. Schliefer, A. and Vishney, R. (1989), 'Industrialisation and the big push', *Journal of Political Economy*, 97, pp. 1003–26.
Nadiri, I. (1993), 'Innovations and technological spillovers', mimeo, Cambridge, Mass., NBER.
Nickell, S. (1996), 'Competition and corporate performance', *Journal of Political Economy*, 104, pp. 724–46.
Nickell, S., Nicolitsas, D. and Dryden, N. (1997), 'What makes firms perform well?', *European Economic Review*, 41, pp. 783–96.
Nickell, S., Wadhwani, S. and Wall, M. (1992), 'Productivity growth in UK companies: 1975-1986', *European Economic Review*, 36, pp. 1055–91.
Oulton, N. (1996), 'Increasing returns and externalities in UK manufacturing: myth or reality?', *Journal of Industrial Economics*, 44, pp. 99–113.
Oulton, N. and O'Mahony, M. (1994), *Productivity and Growth*, Cambridge, Cambridge University Press.
Porter, M. (1990), *The Competitive Advantage of Nations*, London, Macmillan.
Roeger, W. (1995), 'Can imperfect competition explain the difference between primal and dual productivity measures? Estimates for US manufacturing', *Journal of Political Economy*, 103, pp. 316–30.
Saxenian, A. (1994), *Regional Advantage*, Cambridge Mass., Harvard University Press.
Scherer, F.M. (1982), 'Inter-industry technology flows and productivity growth', *Review of Economics and Statistics*, 64, pp. 627–34.
Schliefer, A. (1986), 'Implementation cycles', *Journal of Political Economy*, 94, pp. 1163–90.
Street, J., Carroll, R. and Ruppert, D. (1988), 'A note on computing robust regression estimates via iteratively reweighted least squares', *The American Statistician*, 42, pp. 152–4.
Wadhwani, S. and Wall, M. (1986), 'The UK capital stock: new estimates of premature scrapping', *Oxford Review of Economic Policy*, 2, pp. 44–55.
Wakelin, K. (1997), 'Productivity growth and R&D expenditure in UK manufacturing firms', mimeo, MERIT, University of Maastricht.

11 Human capital, investment and innovation: what are the connections?

STEPHEN NICKELL AND DAPHNE NICOLITSAS[1]

1 Introduction

In 1988, Finegold and Soskice put forward the interesting idea that it is possible for a country to be trapped in a 'low skills' equilibrium. One interpretation of this notion is that the incentives to accumulate human capital depend positively on the stock of physical and/or knowledge capital, and the incentives to accumulate these latter forms of capital depend on the stock of human capital. Thus both forms of investment, human capital and physical/knowledge capital, may be strategic complements, which opens up the possibility of multiple equilibria.

Perhaps the most satisfactory formalisation of this notion is presented in Redding (1996), who embeds everything into an endogenous growth framework, although Acemoglu (1996) and Snower (1994) had previously provided models with some of the relevant elements. In the Redding model, the economy's equilibrium rate of growth depends on both the rate of R&D expenditure and the rate of human capital accumulation. Both forms of investment exhibit pecuniary externalities and are strategic complements. In the presence of indivisibilities in either the R&D or the human capital accumulation technology, multiple equilibria exist and one such is readily interpreted as having a 'low-skills' property.

Our purpose here is to look at one side of this story and investigate whether or not there is any connection between the availability of human capital and the rates of accumulation of physical capital (fixed capital investment) or knowledge capital (R&D expenditure). We already have evidence which is tangentially related to this question, notably on capital-skill complementarity. Furthermore Bartel and Lichtenberg (1987), on the basis of an analysis of the demand for labour by skill, conclude that educated workers have a comparative advantage in the implementation of new technologies. Then there is a substantial body of evidence that new technologies lead to increases in the demand for, and/or the pay of, skilled workers (see, for example, Berman, Bound and Griliches, 1994, Doms, Dune and Troske, 1997, Autor, Katz and Krueger, 1997, Bell, 1996, Machin, Ryan and Van Reenan, 1996). However, while all this evidence is suggestive, none is directly concerned with the investment side of the story.

In the next section we discuss the theoretical background and empirical formulation of our models. This is followed by a presentation of our results where we explain fixed capital investment and R&D expenditures at the firm level. We conclude with a brief summary.

2 Theoretical and empirical background

Although expenditures on R&D and physical capital both can be thought of as investments, the nature of these expenditures is entirely different. Approximately 90 per cent of R&D expenditure goes on the pay of more or less skilled employees plus the materials necessary for them to do their work. Fixed investment expenditure, on the other hand, is entirely spent on capital goods.

The different nature of these expenditures explains the well-known fact that fixed capital investment expenditures exhibit far more variability and far less persistence than R&D expenditures (see, for example, Lach and Schankerman, 1989, or Mairesse and Siu, 1984). The argument is straightforward. The cost of adjusting expenditure on R&D depends on the cost of adjusting the number of skilled employees. Since the work of Oi (1962), this is known to be very high. The cost of adjusting expenditure on fixed capital merely involves the relatively minor costs of adjusting investment plans. It is important to recognise that the cost of adjusting investment expenditures is not the same as the cost of adjusting the capital stock (K). The former is the cost of adjusting ΔK, which is small, the latter is the cost of adjusting K, which is large.[2]

Because of the different nature of the expenditures, we need a different model for fixed capital and R&D expenditures. We start with the former, because the theory is well developed.

2.1 An investment model

The model we use here is that set out in Nickell and Nicolitsas (1997). It is based on the standard quadratic adjustment cost framework, using the stock version of the first order condition with explicit modelling of the current and future marginal returns to capital. We eschew the use of the Q and Euler equation versions of this model essentially because they have been shown to be unsatisfactory in much previous work (see Nickell and Nicolitsas, 1997).[3]

Consider a firm which generates a real net earnings stream, $\pi(K(t),t)$ and the total flow cost of purchasing a stream of new capital goods, $I(t)$, is given by $C(I(t),t)$. This includes both the adjustment costs and the purchase price, and is strictly convex in I (that is, $C_{11} > 0$). If one unit of new capital is purchased at time t, this costs $C_1(I(t),t)$ and generates an additional revenue stream $e^{-\delta s}\pi_K(K(t+s),t+s)$ from $s = 0$ to ∞. The parameter δ is the fixed rate at which capital decays. Investment at t is set at the point where the total cost of an additional unit just balances the (expected) present value of the additional revenue stream, using a real discount rate r. Thus

$$C_1(I(t),t) = \int_0^\infty e^{-(r+\delta)s} \pi_K(K(t+s),t+s)ds \qquad (1)$$

In order to pursue this model, we must be explicit about the real net earnings stream of the firm. To simplify the analysis, we ignore the firm's R&D expenditures, supposing that decisions about R&D are taken prior to all the decisions described here. This is an innocuous assumption provided there is no feedback from investment to R&D, which is consistent with results reported below and in Mairesse and Siu (1984), Lach and Schankerman (1989), and Nickell and Nicolitsas (1997).

Suppose the firm has a constant returns production function of the form $AF(K,N_1,N_2)$ and faces a demand curve $P^{-\eta}$ where A is the total factor productivity coefficient, N_1 is skilled employment, N_2 is unskilled employment and P is the real price of output (normalised on the price of GDP). Supposing first there are no employment adjustment costs, then given skilled and unskilled real wages (normalised on the price of GDP), the firm will choose N_1, N_2 and P at each point to solve

$$\max_{P,N_1,N_2} \quad PAF(K,N_1,N_2) - W_1 N_1 - W_2 N_2$$

subject to $AF(K,N_1,N_2) = P^{-\eta}$

The solution to this problem has the form

$$N_i^* / K = g^i(W_1 / MA, W_2 / MA) \qquad (2)$$

where $M = P[1-\frac{1}{\eta}]$ = marginal revenue. So the skilled and unskilled employment levels will satisfy (2) at each point on the optimal path.

Turning to the determination of investment, we must first be more explicit about the function C, which measures the total cost of purchasing and installing an investment stream $I(t)$. We suppose that C is strictly convex in I and also depends positively on the real price of capital goods, $P^I(t)$, and negatively on the scale of the firm, T.[4] So the net earnings stream π is given by

$$\pi = PAF(K, N_1^*, N_2^*) - W_1 N_1^* - W_1 N_2^* - C(I, P^I, T)$$

Using the envelope theorem, marginal net earnings, π_K, satisfy

$$\pi_K = MAF_K(1, N_1^*/K, N_2^*/K) \qquad \text{(constant returns)}$$

Finally, using (2), we can rewrite π_K as

$$\pi_K = MAg(W_1/MA, W_2/MA)$$

or

$$\pi_K = MAf(W/MA, W_1/W_2) \qquad (3)$$

where W is the average wage in the firm. We prefer the last form of the equation because, while we can observe W at the firm level, we only have W_1, W_2 at the industry level. In a standard model with some degree of complementarity between capital and skilled labour, we would expect to see $f_1 < 0$, $f_2 < 0$. Ultimately, of course, this remains an empirical question.

We may now use (1) and (3) to write down an investment rule

$$C_1(I(t), P^I(t), T) = E_t \int_0^\infty e^{-(r(t)+\delta)} M(t+s) A(t+s) f\left[\frac{W(t+s)}{M(t+s)A(t+s)}, \frac{W_1(t+s)}{W_2(t+s)}\right] ds \quad (4)$$

which includes the expectations operator to emphasise that future values, M, A, W, W_1, W_2 are unknown at time t.

It is immediately clear from (4) how the availability of human capital influences investment. If the relative wage of skilled labour has a negative impact on the marginal product of capital, then an increase in the stock of human capital, which will result in a lower relative wage for skilled labour, will lead to a higher current and expected marginal product of capital. However, it is worth pursuing this link a stage further because we have information not only on relative wages but on so-called skill shortages, which may directly reflect changes in relative quantities of labour, at least relative to the demand for labour.

Suppose, for one reason or another, the firm is unable to attain the optimal levels of employment N_1^*, N_2^* at all times. Then using a first order expansion, we can extend equation (3) to

$$\pi_K = MAf(W/MA, W_1/W_2) - \pi_{KN_1}(N_1^* - N_1) - \pi_{KN_2}(N_2^* - N_2) \quad (5)$$

where $(N^* - N)$ is the extent to which the actual value of employment, N, falls short of the optimal level. We can define this shortfall as the shortage of each type of labour, S_1, S_2. That is $S_1 = (N_1^* - N_1)$, $S_2 = (N_2^* - N_2)$. Thus we can add to the right-hand side of equation (4), a further term of the form

$$E_t \int_0^\infty e^{-(r(t)+\delta)s} [-\pi_{KN_1} S_1(t+s) - \pi_{KN_2} S_2(t+s)] ds \quad (6)$$

Note that we would typically expect $\pi_{KN_1} > 0, \pi_{KN_2} < 0$, that is the marginal product of capital increases in skilled labour and decreases in unskilled labour. So skilled (unskilled) labour shortages would then have a negative (positive) impact on investment. Again, however, this is ultimately an empirical matter.

In order to translate (4) + (6) into a form suitable for empirical analysis, we need explicit functional forms for C_1 and f. These we take to be

$$C_1 = B_1 I(t)^{b_1} P^I(t)^{b_2} T; \quad f = B_2 x_1^{-b_3} x_2^{-b_4}.$$

Furthermore, we suppose that the future variables $M(t+s)$, $A(t+s)$, $W(t+s)$, $W_1(t+s)/W_2(t+s)$, $S_1(t+s)$, $S_2(t+s)$ can be written as $M(t+s) = M(t)e^{g_m s}$; $A(t+s) = A(t)e^{g_a s}$; $W(t+s) = W(t)e^{g_w s}$, $W_1(t+s)/W_2(t+s) = W_1(t)/W_2(t)$, $S_i(t+s) = S_i(t)$, $i = 1,2$, where g_m, g_a, g_w are expected growth rates. Note that we are supposing that expectations about the relative price of skilled labour and skilled labour shortages are static.

Then assuming that the expected growth rates are small relative to the total discount factor, $(r+\delta)$, and that the labour shortage terms have a small effect relative to the remainder, we can rewrite (4) + (6) as[5]

$$\ln I_t = \alpha_0 - \alpha_1 \ln P_t^I + \alpha_2 \ln M_t + \alpha_3 \ln A_t - \alpha_4 \ln W_t$$
$$- \alpha_5 \ln(W_{1t}/W_{2t}) - \alpha_6 S_{1t} + \alpha_7 S_{2t} + \alpha_8 g_{mt} \quad (7)$$
$$+ \alpha_9 g_{at} - \alpha_{10} g_{wt} - \alpha_{11} r_t$$

This involves linearising around the mean real discount rate, $\bar{r}$. The coefficients are given by

$$\alpha_1 = b_2/b_1, \alpha_2 = \alpha_3 = (1+b_3)/b_1, \alpha_4 = b_3/b_1, \alpha_5 = b_4/b_1$$
$$\alpha_6 = \pi_{KN_1}/b_1 \bar{\chi}(\bar{r}+\delta), \alpha_7 = -\pi_{KN_2}/b_1 \bar{\chi}(\bar{r}+\delta)$$
$$\alpha_8 = \alpha_9 = (1+b_3)/b_1(\bar{r}+\delta), \alpha_{10} = b_3/b_1(\bar{r}+\delta), \alpha_{11} > 0,$$

where $\bar{\chi}$ is the mean of the integral on the right of (4).

This approximation relies on gross investment remaining positive and on the scale factor being absorbed into the constant term which must, consequently, be firm specific. Most importantly it is flexible enough to enable us to analyse the impact of human capital on investment while properly controlling for other relevant factors.

2.2 An R&D model

While the theoretical foundations of the determinants of fixed capital investment are both relatively straightforward and highly developed, the same cannot be said for *R&D*. Rigorous theoretical models tend to be highly abstract as, for example, in the patent race literature (see Tirole, 1988, chapter 10, for example). As a consequence, empirical models typically consist of little more than a list of judiciously chosen explanatory variables with some theoretically based comments on their expected effects. Our approach will be no different.

We begin by discussing the essential subject matter of this chapter, namely the role of human capital. It is almost a truism that there will be more R&D in firms which employ more skilled workers (see Menezes-Filho, Ulph and Van Reenan, 1995, for confirmation of this fact). But this does not tell us that were skilled workers to become less plentiful and more expensive, less R&D would be undertaken.[6] In order to provide convincing evidence in favour of this hypothesis, we need to demonstrate that reductions in R&D expenditure actually occur *in response* to rises in the price and/or decreases in the availability of skilled workers. This suggests that we must focus on the time series variation in the data, and given the high adjustment costs associated with R&D expenditure, a relatively long time-period will probably be required.

Concerning other variables which might influence R&D expenditure, here we shall follow the empirical literature (see, for example, Lach and Schankerman, 1989, Hall, 1992, Himmelberg and Petersen, 1994). The variables we use include investment in fixed capital, real sales and some measure of financial pressure. This last is thought to be particularly important in the R&D context because, for obvious reasons, informational asymmetries are particularly prominent in R&D activities. On the other hand, we shall eschew the use of an average Q type variable because any measured impact is so hard to interpret in a world where investment announcements have a big impact on equity prices and hence Q (see McConnell and Muscarella, 1985, for example). So the basic R&D model will have the structure

$$\ln RD_t = \beta_0 + \beta_1 \ln Y_t - \beta_2 FP_t + \beta_3 \ln I_t - \beta_4 \ln(W_{1t}/W_{2t}) - \beta_5 S_{1t} + \beta_6 S_{2t} \quad (8)$$

where RD = real R&D expenditure, Y = real sales, FP = financial pressure, I = fixed capital investment, W_1/W_2 = relative wages of skilled workers and $S_1(S_2)$ = shortage of skilled (unskilled) labour. The constant term, β_0, will be firm specific to capture stable firm characteristics such as scale and the variables are lagged because of the time taken for exogenous shifts to impact on a variable which is as sticky as R&D expenditure.

3 Data, empirical formulation and results

3.1 The data

Our basic data set consists of the company accounts, as reported by EXSTAT/EXTEL, for those manufacturing companies which report R&D expenditures. It covers the period 1972 to 1994 and refers to the 100 or so companies which have six or more consecutive years of data. Until 1989, UK companies were not obliged to disclose R&D expenditure in their Profit and Loss Account. However, a small number of major R&D spenders have done so for many years and after 1989, all large companies reported the relevant figures. It is the reporting of R&D which generates the vast majority of the variation of the number of firms in the sample (see appendix tables 11.A1 and 11.A2). Since we always include firm fixed effects in our models, these variations in data reporting are unlikely to cause serious sample selection biases. We are essentially focusing on the time series variation in the data.

The particular variables we use in our empirical investigation are described in detail in the Data Appendix. For each company they include real gross investment, where we normalise on a price index of plant and machinery, and real R&D expenditure where the deflator is the industry specific R&D price index reported in Cameron (1996). We also use real sales, using an industry output price deflator, and average wages, which are total labour costs divided by employment. Employment is simply the number of employees whereas the real capital stock requires a complex computation described in Wadhwani and Wall (1986). The wages of skilled relative to unskilled workers are available at the industry level, and are based on the average weekly earnings of professional and related workers in science and technology divided by the average weekly earnings of manual workers. Labour shortages are another industry level variable measured by the proportion of firms whose output will be constrained by a shortage of skilled (unskilled) labour over the subsequent four months. We also make use of various financial variables described later as well as an industry based weighted average measure of real R&D expenditures in those industries which supply the industry in which the firm is located.

3.2 The empirical model (investment)

We start from equation (7). In the following, i is a firm subscript and j is the subscript for the industry in which the firm is located. Specific points of model specification are as follows.
(i) All explanatory variables are lagged one period. Investment decisions are taken prior to the purchase and installation of new capital goods and the gap between decision and expenditure is such that a 1-year lag is not unreasonable (see Nickell, 1978, p. 289 for survey evidence on this issue).
(ii) The method of computing the total factor productivity term, A, is crucial. This will be based on the Solow residual which must be properly smoothed to remove the impact of unmeasured fluctuations in the intensity of use of factors. Furthermore, it is useful if we can actually incorporate the information on the accumulation of knowledge capital contained in R&D expenditure in our measure of total factor productivity. For example, suppose total factor productivity, A_{it}, depends on knowledge capital, R_{it}, via the constant elasticity form

$$A_{it} = A_i R_{it}^{\gamma_j}$$

Suppose, further, that the firms' R&D, RD_{it}, and the suppliers' R&D, RDS_{jt}, both contribute to knowledge capital with a lag, so we have

$$R_{it} = RD_{it-1} + \omega_j RDS_{jt-1} + (1-\delta) R_{it-1}$$

The role of suppliers' R&D here arises because if the goods purchased by a firm embody innovations, learning how to work with these new goods may enhance the knowledge base of the company. It is now straightforward to derive the log-linear approximation

$$\ln A_{it} - (1-\delta) \ln A_{it-1} = \beta_i + \beta_{1j} \ln RD_{it-1} + \beta_{3j} \ln RDS_{jt-1} \qquad (9)$$

This analysis suggests the following procedure. First, for each company, compute the Solow residual,

$$\Delta \ln \tilde{A}_{it} = \Delta \ln Y_{it} - \alpha_i \Delta \ln N_{it} - (1-\alpha_i) \Delta \ln K_{it} \qquad (10)$$

with $\ln \tilde{A}_{i0} = 0$. Y = real sales, N = employment, K = capital stock and $\alpha = 1.4 \times$ average wage share in value added (1.4 is an estimate of the ratio of price to marginal cost based on arguments provided in Bean and

Symons, 1989). Experimenting with this number had a marginal impact on the final results.

Second, for each 2-digit industry, we pool the firms within that industry and run a regression based on a generalisation of (9), namely

$$\ln \tilde{A}_{it} - (1-\delta)\ln \tilde{A}_{it-1} = \beta_i + \beta_{1j} \ln RD_{it-1} + \beta_{2j} \ln RD_{it-2} \\ + \beta_{3j} \ln RDS_{jt-1} + \beta_{4j} \ln RDS_{jt-2} + \beta_{5j} OH_{jt} \\ + \beta_{6j}(OH_{jt})^{-1} + \beta_{7i} t \quad (11)$$

OH is overtime hours divided by standard hours and the role of OH, $(OH)^{-1}$ is to control for cyclical fluctuations, the non-linear form being suggested by Muellbauer (1984). Then we define trend total factor productivity, A_{it}, by

$$\ln A_{it} - (1-\delta)\ln A_{it-1} = \hat{\beta}_i + \hat{\beta}_{1j} \ln RD_{it-1} + \hat{\beta}_{2j} \ln RD_{it-2} \\ + \hat{\beta}_{3j} \ln RDS_{jt-1} + \hat{\beta}_{4j} \ln RDS_{jt-2} + \hat{\beta}_{5j} \overline{OH}_j \\ + \hat{\beta}_{6j}(\overline{OH}_{j.})^{-1} + \beta_{7i} t \quad (12)$$

Note that we set the cycle terms to their means in (12) in order to capture trend productivity.

The estimates of equation (11) are presented in table 11.1. For each industry we estimate two versions, one with $\delta = 0.4$ (see Blundell, Griffith and Van Reenan, 1995, for example) and one where $\delta = 1$, where we attempt to capture the long-run impact of R&D on total factor productivity. The R&D terms generally have an overall positive impact, in the form of a net positive effect in levels or first differences. The coefficients are not very well determined but it should be recognised that each firm has its own constant and trend, so the equations are heavily parameterised. Our primary purpose is to estimate trend productivity, not to explain it in detail.

(iii) The next problem is to specify the real marginal revenue, M. We do not have firm specific output prices, only those referring to the relevant 3 digit industry. Since we feel it is easier to model the firm's marginal revenue relative to the industry price, we first note that

$$\ln M_{it} = \ln M_{it}/P_{jt} + \ln P_{jt}; \quad g_{imt} = g_{i\tilde{m}} + g_{pt}$$

where $g_{\tilde{m}}$ is the expected growth of M_j/P_j and g_p is the expected growth of the real industry output price, P_j. We then argue that the firm's marginal revenue relative to the industry price will depend positively on demand for the firm's output, perhaps negatively on demand for the

Table 11.1 *Regressions to explain the Solow residual (equation (11))*

Dependent variable: $\ln \tilde{A}_{it}$

Independent variables	1		2		3	
	$\delta = 0.4$	$\delta = 1$	$\delta = 0.4$	$\delta = 1$	$\delta = 0.4$	$\delta = 1$
$\ln RD_{it-1}$	0.103	0.085				
	(2.2)	(1.8)				
$\ln RD_{it-2}$	0.103	0.085	0.073	0.045	0.081	0.026
	(2.2)	(1.8)	(1.1)	(0.8)	(0.7)	(0.3)
$\ln RDS_{jt-1}$	0.22	0.60				
	(1.0)	(2.6)				
$\ln RDS_{jt-2}$	0.22	0.60				
	(1.0)	(2.6)				
OH_{jt}	0.61	−1.58	2.34	−1.73	−5.64	−3.56
	(0.1)	(0.2)	(0.4)	(0.3)	(0.7)	(0.5)
$(OH_{jt})^{-1}$	0.054	0.55	0.012	−0.016	−0.006	−0.00
	(0.5)	(0.5)	(0.5)	(0.7)	(0.1)	(0.0)
N	14		15		9	
NT	76		79		49	

Independent variables	4		5		6	
	$\delta = 0.4$	$\delta = 1$	$\delta = 0.4$	$\delta = 1$	$\delta = 0.4$	$\delta = 1$
$\ln RD_{it-1}$	0.014	0.031	0.032	0.26	0.11	0.17
	(0.7)	(1.5)	(0.1)	(1.7)	(1.1)	(2.2)
$\ln RD_{it-2}$	−0.014	−0.031			0.11	0.17
	(0.7)	(1.5)			(1.1)	(2.2)
$\ln RDS_{jt-1}$	−0.56	0.042			0.14	0.11
	(0.8)	(0.6)			(0.3)	(0.3)
$\ln RDS_{jt-2}$	0.056	−0.042	2.05	3.29	−0.14	−0.11
	(0.8)	(0.6)	(1.3)	(4.2)	(0.3)	(0.3)
OH_{jt}	1.07	1.44	4.83	6.74	4.06	4.48
	(1.5)	(2.1)	(0.9)	(2.6)	(0.7)	(0.9)
$(OH_{jt})^{-1}$	0.003	0.002	0.013	0.015	0.053	0.033
	(1.1)	(0.8)	(1.4)	(3.3)	(1.1)	(0.9)
N	67		5		12	
NT	391		20		68	

Notes:

1 = Food, drink, tobacco; 2 = General chemicals; 3 = Pharmaceuticals; 4 = Engineering, vehicles and metal goods; 5 = Textiles, clothing and footwear; 6 = Bricks, pottery, glass, cement, paper, printing and publishing.
The equations are estimated with firm fixed effects and firm specific time trends.

industry output and positively on the firm's R&D insofar as it generates new products (the process innovation part of R&D goes into A).[7] So we shall proxy M_i/P_j using company sales, Y_i, industry sales, Y_j, and RD_i. Thus we rewrite $\alpha_2 \ln M_{it}$, $\alpha_8 g_{imt}$ in (7) as

$$\alpha_2 \ln M_{it} = \alpha_{21} \ln Y_{it} - \alpha_{22} \ln Y_{jt} + \alpha_{23} \ln RD_{it} + \alpha_2 \ln P_{jt}$$

$$\alpha_8 g_{imt} = \alpha_{81} g_{iyt} - \alpha_{82} g_{jyt} + \alpha_8 g_{jpt}$$

We omit the expected growth rate of R&D because we feel that any effects of such a variable can be adequately captured by lags on the RD variable itself.

(iv) The expected growth terms, $g_{iy}, g_{jy}, g_w, g_a, g_p$ are generated using subsidiary regressions. Our strategy for any firm variable z_{it} is to pool firms in each two digit industry and run a regression for each industry of the form

$$\Delta_2 \ln z_{it+2} = \gamma_0 + \gamma_1 \Delta \ln z_{it} + \gamma_2 \Delta \ln z_{it-1} + \sum_K \gamma_{3k} x_{kit}$$

where x are variables which may help to forecast z. Then $g_{izt} = \frac{1}{2}\Delta_2 \ln z_{it+2}$, i.e. the fitted value from the above regression. For an industry variable such as P_j, we simply use separate time series regressions for each industry. The x variables are as follows. For Y_{it}, x contains industry sales; for Y_{jt}, P_{jt}, no extra variables (i.e. equations only contain lags of the dependent variable). For W_{it}/P_{jt}, x contains interest payments/cash flow, profits per employee, aggregate unemployment rate. For A_{it}, x contains R&D expenditure and suppliers' R&D expenditure.

(v) The discount factor, r_{it}, we divide into two parts. The first is the aggregate post-tax real interest rate, r_t, which will, in practice, be absorbed into time dummies. The second is a firm specific premium on borrowing costs which depends on the ratio of debt to 'collateraliseable' net worth. A sensible method of capturing this ratio is to use total interest payments normalised on cash flow. This is a flow equivalent to the ratio of debt to the present value of profits (including interest payments) which does not suffer from the measurement problems associated with stock measures such as the debt to equity ratio.[8] Indeed the inverse of interest payments over cash flow is the level of interest cover which is a measure of leverage used by bankers in assessing the credit worthiness of companies. So in our final equation, we specify the discount factor term as

$$\alpha_{11} r_{it} = \alpha_{11} r_t + \alpha_{111} FP_{it}$$

where FP (financial pressure) = interest payments/cash flow.

If we now rewrite equation (7) to include all the changes described above and add in firm and industry subscripts, we obtain

$$\ln I_{it} = \alpha_{i0} - (\alpha_1 \ln P_t^I + \alpha_{11} r_t) + \alpha_{21} \ln Y_{it-1} - \alpha_{22} \ln Y_{jt-1}$$
$$+ \alpha_{23} \ln RD_{it-1} + (\alpha_2 - \alpha_4) \ln P_{jt-1} + \alpha_3 \ln A_{it-1}$$
$$- \alpha_4 \ln(W_{it-1}/P_{jt-1}) - \alpha_5 \ln(W_{1jt-2}/W_{2jt-2}) - \alpha_6 S_{1jt-2}$$
$$+ \alpha_7 S_{2jt-2} + \alpha_{81} g_{iyt-1} - \alpha_{82} g_{jyt-1} \quad (13)$$
$$+ (\alpha_8 - \alpha_{10}) g_{jpt-1} + \alpha_9 g_{iat-1} - \alpha_{10} g_{iwpt-1}$$
$$- \alpha_{111} FP_{it-1}.$$

Note that we have normalised the wage on the industry price and included the corresponding growth term g_{iwp} (the expected growth of W_i/P_j). In (13), the real investment goods price and the aggregate real interest rate will be absorbed into time dummies and α_{i0} captures all stable unobserved factors including the scale of the firm. Finally, we have included relative skilled wages and skilled/unskilled labour shortages with two lags in order to ensure there is no reverse causality problem. That is, the possibility that high investment leads to skill shortages or higher skilled wages. Of course, the reverse causality effect has the opposite sign to that which we are looking for, so if we do find a negative effect we know it is not due to reverse causality.

3.3 The empirical model (R&D)

Since we have no strong theoretical structure to fall back on, we propose simply to use a general autoregression based on equation (8). Thus we shall estimate

$$\ln RD_{it} = \beta_{0i} + \sum_{k=1}^{2} [\beta_{k0} \ln RD_{it-k} + \beta_{k1} \ln Y_{it-k}$$
$$- \beta_{k2} \ln FP_{it-k} + \beta_{k3} \ln I_{it-k} - \beta_{k4} \ln(W_{1t-k}/W_{2t-k}) \quad (14)$$
$$- \beta_{k5} S_{1t-k} + \beta_{k6} S_{2t-k}]$$

Furthermore, because R&D expenditure is likely to be rather stable over time we restrict ourselves to companies for which we have at least ten consecutive years of data.

Table 11.2 *Investment equations (1976–94)*

Dependent variable: $\ln I_{it}$

	1 $\delta = 0.4$		2 $\delta = 1.0$		3 $\delta = 1.0$	
$\ln Y_{it-1}$	0.12	(0.5)	0.061	(0.2)	0.049	(0.2)
$\ln Y_{jt-1}$	1.25	(1.7)	1.27	(1.7)	1.18	(1.7)
$\ln RD_{it-1}$	0.26	(1.6)	0.27	(1.7)	0.29	(1.8)
$\Delta \ln RDS_{jt-1}$	0.61	(1.3)	0.61	(1.3)	0.49	(1.2)
$\ln A_{it-1}$	1.04	(1.8)	1.17	(2.2)	1.15	(2.2)
$\ln P_{jt-1}$	0.13	(0.1)	0.17	(0.1)	0.29	(0.2)
$\ln(W_i/P_j)_{t-1}$	0.04	(0.1)	0.08	(0.3)	0.035	(0.1)
FP_{it-1}	−0.42	(2.1)	−0.44	(2.2)	−0.46	(2.3)
$\ln(W_{1j}/P_{2j})_{t-2}$	−0.60	(0.7)	−0.64	(0.7)		
S_{1jt-2}	−0.99	(2.2)	−0.95	(2.2)	−0.88	(2.1)
g_{iyt-1}	0.48	(2.1)	1.26	(2.5)	1.20	(2.5)
g_{jyt-1}	−4.78	(1.8)	−5.15	(1.9)	−5.74	(2.2)
g_{iat-1}	1.29	(2.1)	1.30	(1.0)	1.34	(1.0)
g_{jpt-1}	6.06	(0.8)	5.80	(0.8)	6.20	(0.9)
g_{iwpt-1}	−0.40	(1.2)	−0.38	(1.2)	−0.39	(1.3)
se	0.63		0.63		0.63	
N	99		99		99	
NT	420		420		420	

Notes:

I_i = real fixed capital investment; RD_i = real expenditure on R&D, Y_i = real sales; Y_j = real industry sales; RDS_j = real expenditure on R&D in supplier industries (weighted average); A_i = total factor productivity; P_j = real price of output; W_i = real wage; FP_i = interest payments/cash flow; W_{1j} = average weekly earnings for professional and related workers in science, engineering and technology; W_{2j} = average weekly earnings for manual workers; S_{1j} = proportion of firms whose output will be constrained by a shortage of skilled labour over the next four months; g_{iy} = expected growth of real sales; g_{jy} = expected growth of industry sales; g_{ia} = expected TFP growth; g_{jp} = expected growth of real output prices; g_{iwp} = expected growth of real product wages; i = firm; j = industry.

Equations are estimated using the standard within estimator. Heteroscedasticity robust t ratios in brackets. Time dummies and unskilled labour shortages are omitted because they are not jointly significant and have no impact on the coefficients except for that on industry output.

3.4 Results

In table 11.2, we report within estimates of the investment equation (13). In column 1, we set δ to 0.4 when defining total factor productivity, A (see equation (12)). In the remaining columns, we set δ = 1.0. The specification corresponds to that in equation (13) with the following exceptions. First, we include the change in suppliers' R&D because this appears to have some role in the results described in Nickell and Nicolitsas (1997). Second, we drop unskilled labour shortages and all the time dummies. They are all individually insignificant and tend to be jointly insignificant at the 1 per cent level. (The test statistics are $F(20,383) = 1.95$, $F(20,383) = 1.91$, $F(21,383) = 1.85$ for the three columns.) Their exclusion makes more or less no difference to the size of any of the coefficients except for that on industry sales. In particular the key coefficient on the skill shortage variable is unaltered.

Before commenting on the parameter estimates, it is worth remarking briefly on endogeneity problems. Although the regressors are all lagged, this will not entirely eliminate these problems because there may be potential order ($1/_T$) biases due to possible correlations between the equation error and current and future regressors. These biases arise via the within-groups transformation.

Of the regressors in the model, those which are potentially endogenous include sales (Y_i), own R&D (RD_i), trend total factor productivity (A_i), wages (W_i), relative skilled wages, (W_{1j}/W_{2j}) and skill shortages (S_{1j}). Considering each of these in turn, there is no evidence that investment shocks influence R&D (see below) so there should be no problems with either RD or A (which is a function only of RD and time trends, see equation (12)). Investment shocks may affect company sales and wages because of increased productive potential but these effects are surely small since a substantial change in annual investment corresponds to only a small change in overall capacity. With regard to relative skilled wages and skill shortages, these are, of course, industry based variables which reduces the effect of firm based shocks. Furthermore, the expected impact of investment shocks is positive which is opposite in sign to the one we are searching for. So if we find a negative effect, it is all the more persuasive.

Turning to the parameter estimates, it is worth noting the positive effect of total factor productivity and the negative one of financial pressure. However the key finding is the significant negative effect of skilled labour shortages. A 10 percentage point rise in the number of companies in the firm's industry reporting a shortage of skilled labour corresponds to a subsequent 10 per cent fall in the firm's investment expenditure. This finding is perhaps more persuasive when it is recalled that increases in skill shortages are associated with booms which typically lead to increased investment expenditures.

We next consider the effect of skill shortages on R&D and so, in table

Table 11.3 *Investment and R&D equations (1976–94)*

Dependent variable	$\ln I_{it}$		$\ln RD_{it}$	
Independent variables				
$\ln I_{it-1}$	0.48	(4.8)	0.006	(0.2)
$\ln I_{it-2}$	−0.16	(3.1)	−0.006	(0.2)
$\ln RD_{it-1}$	−0.038	(0.2)	0.70	(5.0)
$\ln RD_{it-2}$	0.43	(2.7)	−0.087	(0.9)
$\ln Y_{it-1}$	0.51	(1.2)	0.25	(1.6)
$\ln Y_{it-2}$	−0.19	(0.5)	0.059	(0.5)
FP_{it-1}	−0.016	(0.1)	0.051	(0.3)
FP_{it-2}	−0.43	(2.6)	−0.11	(0.8)
S_{1jt-1}	0.20	(0.2)	−0.39	(1.5)
S_{1jt-2}	−0.91	(1.4)	0.37	(2.2)
S_{2jt-1}	1.81	(1.1)	1.62	(2.5)
S_{2jt-2}	0.77	(0.7)	−0.93	(2.2)
se	0.58		0.24	
N	18		18	
NT	170		170	

Notes:
I = real fixed capital investment; RD = real expenditure on R&D; Y = real sales; FP = interest payments/cash flow; $S_1(S_2)$ = proportion of firms whose output will be constrained by a shortage of skilled (other or unskilled) labour over the next four months; i = firm; j = industry.
Equations are estimated using the standard within (fixed effects) estimator. The t ratios are corrected for heteroscedasticity. All equations are estimated using DPD by Arrelano and Bond (1991).

11.3, we report some multivariate autoregressions based on equation (14), although we omit the wage terms because their effect was always negligible. For comparison purposes, we also report a similar equation explaining investment expenditures. These autoregressions are estimated using the standard within procedure and have a time dimension $T \geq 9$. That is why they include only eighteen firms. As we have already noted, we only use this very restricted sample because the long time dimension is required to investigate exogenous effects on a variable which is as sticky as R&D expenditure. It then has the added advantage that the fixed effects (downward) bias on the lagged dependent variable coefficients is not so large as to nullify the whole procedure. Indeed the lagged dependent variable coefficient is probably understated by around 0.15 (see Nickell, 1981).

Looking again at investment, there are significant effects from R&D (positive) and financial pressure (negative) and some indication of a negative skill shortage effect of the same order of magnitude as that shown in table 11.2, although not very significant ($t = 1.4$). For R&D expenditure, things look rather different. First, the overall level of persistence is much higher than for investment (the sum of the lagged dependent variable coefficients for R&D is 0.61 as opposed to 0.32 for I). Second, there is no feedback from investment in fixed capital, nor is there any financial pressure effect. There is a hint of a sales effect but otherwise there seem only to be temporary effects from skilled labour shortages (negative) and unskilled labour shortages (positive). Thus, a permanent 10 percentage point rise in the number of companies in the firm's industry reporting a shortage of skilled labour leads to a subsequent 4 per cent reduction in the firm's R&D expenditures for one period only.

4 Summary and conclusions

It has been suggested by Finegold and Soskice (1988), and more formally by Redding (1996), that it is possible for a country to be trapped in a 'low skills' equilibrium. The idea is that investments in human capital and in physical/knowledge capital are strategic complements because the more you have of one type of capital, the greater the incentive to invest in the other type. The potential for multiple equilibria is clear.

Our purpose here has been to investigate one branch of this strategic complementarity, namely the possibility that a shortage of human capital will reduce the level of fixed capital, and/or R&D, investment expenditure. Using panel data on a large number of manufacturing companies which undertake R&D expenditure, we analyse the *ceteris paribus* effects of survey based reports of skilled labour shortages at the industry level on subsequent investment and R&D expenditures at the firm level. Our results suggest that a permanent 10 percentage point increase in the number of firms reporting skilled labour shortages in the industry to which a firm belongs will lead to a permanent 10 per cent reduction in its fixed capital investment and a temporary 4 per cent reduction in its R&D expenditure. These preliminary results suggest that the theoretical possibility that investment in human capital and physical/knowledge capital are strategic complements is one which merits further empirical investigation.

Data Appendix

A1 The data

The data area taken from the EXSTAT and EXTEL databases. These are based on the published accounts of UK companies and are supplied by Extel Financial Ltd.

Our sample consists of 99 firms with at least 6 consecutive observations in the period 1972–94. Tables 11.A1 and 11.A2 give the number of firms available in each year and the number of firms by number of consecutive observations respectively.

Table 11.A1

Year	No. of firms	Year	No. of firms	Year	No. of firms
1972	3	1980	18	1988	72
1973	4	1981	21	1989	83
1974	5	1982	36	1990	81
1975	5	1983	41	1991	65
1976	7	1984	41	1992	60
1977	10	1985	41	1993	59
1978	11	1986	45	1994	50
1979	15	1987	43		

Table 11.A2

Observations	Firms	Observations	Firms	Observations	Firms
6	28	11	3	16	1
7	30	12	4	17	1
8	11	13	2	19	1
9	12	14	2	20	1
10	2	15	1		

For years prior to 1991 the source is the EXSTAT database deposited at the University of Bath, from where it was accessible to the academic community. For years after 1991 the source is EXTEL Company Analysis. The variation in the sample size is due either to the non-reporting of R&D prior to 1989 or to the problems in matching EXSTAT and EXTEL data after 1990. Firms are not entering or exiting the sample for economic reasons but for data reasons.

A2 Firm specific variables

1. *Investment (I)*. Gross investment defined as the sum of additions to property and other tangible assets plus property and other tangible assets of new subsidiary companies (EXSTAT items CC3+CC5+CC11+C13/ EXTEL items it.pag.aa+it.og.aa). This is normalised on an industry specific price index for plant and machinery - see below.
2. *R&D (RD)*. Research & Development expenditure (EXSTAT item C65/ EXTEL item te2.rsd). This is normalised on an R&D deflator for manufacturing constructed by Gavin Cameron (see Cameron, 1996, chapter 5, data annex 3).
3. *Output (Y)*. Sales turnover (EXSTAT item C31/EXTEL item sl). This is normalised on an industry specific producer price index – see below.
4. *Capital (K)*. This is based on transforming net tangible assets at historic cost into the same variable at current replacement cost and then normalising on the price index for plant and machinery. This is an extremely complex calculation and full details are provided in Wadhwani and Wall (1986).
5. *Financial pressure (FP)*. This is defined as:

$$\frac{\text{Interest payments}}{\text{Profits before tax + Depreciation + Interest payments}}$$

EXSTAT ITEMS $\quad \dfrac{C57}{C34 + C52 + C57}$

EXTEL ITEMS $\quad \dfrac{\text{fc.st + fc.lt + fc.uo}}{\text{pbt + tel.d + fc.st + fc.lt + fc.uo}}$

For those firms for which the denominator in (1) above takes on a negative value, that is for firms facing losses prior to the payment of interest rates and the deduction of depreciation, FP is set equal to 1.

6. *Wage (W)*. Staff costs (EXSTAT item C63/EXTEL item tel.sf) over employment (EXSTAT item C19/EXTEL item nemp). This is normalised on an industry specific producer price index – see below.
7. *Total factor productivity (A)*. First we compute the Solow residual for each firm $\tilde{A}$ defined by

$$\Delta \ln \tilde{A} = \Delta \ln Y - \alpha \Delta \ln N - (1 - \alpha) \Delta \ln K$$

where α is 1.4 times the average for each firm of staff costs (see above) over value added (see below). Setting $\ln \tilde{A}$ for the first period equal to zero we then have for each subsequent period:

$$\ln \tilde{A}_t = \Delta \ln \tilde{A} + \ln \tilde{A}_{t-1}$$

$\ln \tilde{A}$ is then regressed for each industry on R&D expenditure, suppliers' R&D expenditure and overtime hours as described in (11) and table 11.1

in the main text. Then ln A is the fitted value from this regression setting overtime hours to its sample average value.

8 *Value added.* Value added is defined as the sum of staff costs (EXSTAT item C63/EXTEL item tel.sf) + profits before tax (EXSTAT item C34/ EXTEL item pbt) + depreciation (EXSTAT item C52/EXTEL item tel.d) + interest payments (EXSTAT item C53 + C54 + C55/EXTEL items fc.st + fc.it + fc.uo).

9 *Profits per employee* [$\frac{\pi}{N}$] Profits before tax (EXSTAT item C34/EXTEL item pbt) over employment (EXSTAT ote, C19/EXTEL item nemp).

A3 Industry level variables

1 *Prices of plant and machinery.* Price indices for plant and machinery bought as fixed assets by 3-digit industry from *Business Monitors* MO18, MM17 and its successor.

2 *Producer price index* (P_j). Producers' price indices for 3-digit industries from the *Annual Abstract of Statistics* and *Business Monitor* MM22 various issues.

3 *Industry output* (Y_j). Index of real output by 2-digit industry from the *Monthly Digest of Statistics* (table 7.1) various issues.

4 *Industry wages* (W_j). Weekly earnings of full-time manual males from the *Employment Gazette* (table 5.4) prior to 1990 and the *New Earnings Survey* (table 54, part C) for 1991–4.

5 *Research & Development expenditure of suppliers to the industry* (RDS_j). This is a weighted average of industry R&D expenditure where the weights are the coefficients in a technology matrix (α_{jk}) and indicate the share of total supplies to industry (j) produced by industry k. Weights are renormalised to sum to unity after setting α_{jj} to zero. R&D expenditure is from the OECD STAN database and the 6th edition of *Research and Development in UK Business*. R&D has been deflated by a manufacturing R&D deflator. The source for the technology matix is the CSO publication 'Input-Output Tables for the UK, 1984'.

6 *Industry overtime hours* (OH_j). This is the ratio of average weekly overtime hours per operative on overtime multiplied by the proportion of manual workers receiving overtime pay over normal weekly hours of manual workers from the *New Earnings Survey*, part C, tables 54 and 79 various issues.

7 *Skilled labour shortages* (S_{1j}). The proportion of firms whose output over the next four months will be constrained by a shortage of skilled labour. Confederation of British Industry (CBI) *Quarterly Industrial Trends Sur-*

vey. This is based on a large random sample of mainly manufacturing firms in Britain.

8   *Unskilled labour shortages* (S_{2i}). The proportion of firms whose output over the next four months will be constrained by a shortage of other labour. *CBI Quarterly Industrial Trends Survey*.

9   *Wages of skilled workers* (W_{1i}). Weekly earnings for professional occupations. *New Earnings Survey*, table 104, various issues. For 1972, we use group 3: Engineers, Scientists and Technologists. For 1973–90, we use group 5: Professional and Related in Science, Engineering and Technology. For 1991–4, we use group 2; Professional Occupations.

10 *Wages of unskilled workers* (W_{2i}). Weekly earnings for manual occupations. *New Earnings Survey*, table 86, various issues.

Notes

1 This chapter is part of an ESRC funded project on Innovation, Investment and Human Capital and has been carried out under the auspices of the Corporate Performance Programme of the ESRC Centre for Economic Performance.

2 Lach and Schankerman (1989) fail to recognise this distinction when they argue against the view that the costs of adjusting R&D are higher than the costs of adjusting investment. Thus they note that 'notwithstanding the skill level of R&D labor, one would presume that it exhibits less rigidity than physical capital'. Indeed it does. However the correct comparison is with the rigidity of ΔK, not the rigidity of K.

3 The basic problems arise first from the endogeneity of Q and second from the fact that freely estimated Euler equations generally have coefficients which are completely inconsistent with an Euler equation interpretation.

4 We treat the scale of the firm as constant. Over the kind of short time period used here, this is not unreasonable.

5 This involves replacing the log of the expectation of the integral by the expectation of its log, which is only accurate if the variation over time is small.

6 Of course, it is almost inevitable if R&D itself uses substantial numbers of skilled workers. However, we are interested in the bigger effects which would follow from the possibility that the benefits to R&D expenditure are increasing with the availability of human capital.

7 This is not quite correct because when we compute A_i, we use $(P_i Y_i)/P_j$ as our measure of real output. So some of the effect of R&D on A is, in fact, an effect of R&D on P_i/P_j which can arise from product innovations.

8 Note the Debt/PV profits = $\frac{D}{\pi/(r+\rho-g^e)}$ where D is current debt, π = current profits including interest rate, r = risk premium, g^e = expected growth rate of profits. This

ratio can be written as $(r + \rho - g^e)D/\pi$ where $(r + \rho - g^e)D$ is well approximated by current interest payments which will depend on the interest rate and the overall level of risk (positively related to ρ, negatively to g^e).

References

Acemoglu, D. (1996), 'A microfoundation for social increasing returns in capital accumulation', *Quarterly Journal of Economics*, 111, pp. 779–804.

Arrelano, M. and Bond, S. (1991), 'Some tests of specification for panel data', *Review of Economic Studies*, 58, pp. 277–98.

Autor, D.H., Katz, L.F. and Krueger, A.B. (1997), 'Computing inequality: have computers changed the labor market', NBER Working Paper No. 5956, March.

Bartel, A.P. and Lichtenberg, F.R. (1987), 'The comparative advantage of educated workers in implementing new technology', *Review of Economics and Statistics*, 69, pp. 1–11.

Bean, C. and Symons, J. (1989), 'Ten years of Mrs. T.', in Blanchard, O. and Fischer, S. (eds.), *NBER Macroeconomics Annual*, Cambridge Mass., MIT Press.

Bell, B.D. (1996), 'Skill-biased technical change and wages: evidence from a longitudinal data set', Leverhulme Trust Programme, Discussion Paper No. 8, Oxford, Institute of Economics and Statistics.

Berman, E., Bound, J. and Griliches, Z. (1994), 'Changes in the demand for skilled labor within US manufacturing industries: evidence from the Annual Survey of Manufacturing', *Quarterly Journal of Economics*, 109, pp. 367–97.

Blundell, R., Griffith, R. and Van Reenen, J. (1995), 'Market dominance, market value and innovation in a panel of British manufacturing firms', mimeo, Institute for Fiscal Studies.

Cameron, G. (1996), 'Innovation and Economic Growth', Oxford University D.Phil. Thesis, Nuffield College.

Doms, M., Dunne, T. and Troske, K. (1997), 'Workers, wages and technology', *Quarterly Journal of Economics*, 112.

Finegold, D. and Soskice, D. (1988), 'The failure of training in Britain: analysis and prescription', *Oxford Review of Economic Policy*, 4, 3, pp. 21–53.

Hall, B. (1992), 'Investment and R&D at the firm level: does the source of financing matter?', NBER Working Paper No. 4096, June.

Himmelberg, C.P. and Petersen, B.C. (1994), 'R&D and internal finance: a panel study of small firms in high-tech industries', *Review of Economics and Statistics*, 76, pp. 38–51.

Lach, S. and Schankerman, M. (1989), 'Dynamics of R&D and investment in the scientific sector', *Journal of Political Economy*, 97, pp. 880–904.

Machin, S., Ryan, A. and Van Reenen, J. (1996), 'Technology and changes in skill structure: evidence from an international panel of industries', Institute for Fiscal Studies, Working Paper No. W96/6.

Mairesse, J. and Siu, A.K. (1984), 'An extended accelerator model of R&D and physical investment', in Griliches, Z. (ed.), *R&D, Patents and Productivity*, Chicago, University of Chicago Press, pp. 271–97.

McConnell, J.J. and Muscarella, C.J. (1985), 'Corporate capital expenditures and the market value of the firm', *Journal of Financial Economics*, 14, pp. 399–422.

Menezes-Filho, N., Ulph, D. and Van Reenen, J. (1995), 'R&D and union bargaining: evidence from British companies and establishments', mimeo, University College, London, April.

Muellbauer, J. (1984), 'Aggregate production functions and productivity: a new look', CEPR Discussion paper No. 34, London.

Nickell, S.J. (1978), *The Investment Decisions of Firms*, Cambridge, Cambridge University Press.

(1981), 'Biases in dynamic models with fixed effects', *Econometrica*, 49, pp. 1417–26.

Nickell, S. and Nicolitsas, D. (1997), 'Does innovation encourage investment in fixed capital?', mimeo, Institute of Economics and Statistics, University of Oxford.

Oi, W. (1962), 'Labour as a quasi-fixed factor', *Journal of Political Economy*, 70, pp. 538–55.

Redding, S. (1996), 'The low-skill, low-quality trap: strategic complementarities between human capital and R&D', *Economic Journal (RES Conference Volume)*, 106, pp. 458–70.

Snower, D. (1994), 'The low-skill, bad-job trap', CEPR Discussion Paper No. 999, September.

Tirole, J. (1988), *The Theory of Industrial Organization*, Cambridge, Mass., MIT Press.

Wadhwani, S.B. and Wall, M. (1986), 'The UK capital stock – new estimates of premature scrapping', *Oxford Review of Economic Policy*, 2, pp. 44–55.

www.ingramcontent.com/pod-product-compliance
Ingram Content Group UK Ltd.
Pitfield, Milton Keynes, MK11 3LW, UK
UKHW032104190125
453752UK00004B/11